1 MONTH OF
FREE
READING

at

www.ForgottenBooks.com

By purchasing this book you are eligible for one month membership to ForgottenBooks.com, giving you unlimited access to our entire collection of over 1,000,000 titles via our web site and mobile apps.

To claim your free month visit: www.forgottenbooks.com/free908210

ISBN 978-0-266-90918-7
PIBN 10908210

Historic, archived document

Do not assume content reflects current
scientific knowledge, policies, or practices

BULLETIN OF THE U.S. DEPARTMENT OF AGRICULTURE

No. 55

Contribution from the Forest Service, Henry S. Graves, Forester.
March 25, 1914.

(PROFESSIONAL PAPER.)

BALSAM FIR.

By RAPHAEL ZON,
Chief of Forest Investigations.

INTRODUCTION.

The enormous expansion of the pulp industry in this country during the last two decades, with its present annual demand for not less than three and a quarter million cords of coniferous wood, has stimulated the use of balsam fir, which but a few years ago was considered of little value. With the increase in the price of spruce for pulpwood, balsam fir has begun to take its place for rough lumber, laths, shingles, and box shooks. The cutting of balsam fir to any extent for pulp or lumber began only about 20 years ago, as the more valuable species of the northern forests became scarce and as its suitability for many purposes for which only white pine or spruce were originally used became recognized.

Balsam fir, though in general inferior to white pine and red spruce, is now a tree of considerable economic importance in the northeastern forests. It constitutes numerically about 20 per cent of the coniferous forests in northern New York and Maine, and is abundant in many parts of New Hampshire, Vermont, and in the swamps of northern Michigan, northern Wisconsin, and Minnesota. Through prolific seeding and rapid growth it readily reforests cut-over areas and attains sizes suitable for pulpwood in a short time.

The uses for which balsam fir is suited and the appearance of barked wood, especially after it has remained for any length of time in water, are so much like those of spruce that it is commonly sold in mixture with and under the name of spruce, because of a lingering prejudice against balsam fir among pulp manufacturers and lumbermen. This prejudice, formed at the time of still abundant supplies of spruce and

NOTE.—This bulletin deals with all aspects of balsam fir, its distribution, the forest types in which it occurs, the present stand and cut, its economic importance, especially in relation to the paper-pulp industry, methods and cost of lumbering, life history of the tree, characteristics of the wood, rate of growth and yield, and proper methods of management. Balsam fir is found in commercial quantities in the northeastern border States from Maine to Minnesota.

white pine, is based partly on the actual inferiority of balsam fir to those species and partly to insufficient familiarity with the wood.

To determine impartially the economic value of balsam fir, its distribution, present stand and cut in the various States where it occurs, as well as its qualities and possibilities as a forest tree, was the purpose of two summers' study in the Adirondacks, in Maine, and throughout the whole of the tree's commercial range. It was believed that by pointing out the possibility of using balsam fir in places where originally only spruce had been used, and by learning its peculiarities as a forest tree, the heavy drain upon our waning supplies of spruce might be slightly decreased, and that suggestions for the proper management of our spruce forests, in which balsam fir holds an important place, could be formulated.

DISTRIBUTION OF BALSAM FIR.

Balsam fir (*Abies balsamea* Mill.) is a tree chiefly of the Northeast, although it occurs here and there in the mountain ridges of southern Virginia and extends westward in Canada as far as Mackenzie River. (See map, fig. 1.)

Moisture and temperature are the two main factors influencing its distribution. It requires a cold climate and a constant supply of moisture at its roots. A mean annual temperature not exceeding 40° F., with an average summer temperature of not more than 70° F., and a mean annual precipitation of not less than 25 inches evenly distributed throughout the year, are the necessary conditions for its growth. It extends farther north than red spruce, but is left slightly behind by black and white spruce, tamarack, aspen, and paper birch.

Though in Canada balsam fir extends almost to the Rocky Mountains, in which it is doubtless supplanted by Alpine fir (*Abies lasiocarpa*),[1] it does not occur in continuous large forests west of the one hundredth meridian, and in the United States its western limit is found in Minnesota. One of the principal reasons for this is the increasing dryness of the air which the tree encounters in its westerly distribution. The mean annual rainfall gradually decreases from the east toward the west. In Maine, where balsam fir reaches its best development, the rainfall amounts to 43 inches; in Minnesota, where balsam is of poor development, it is less than 26 inches. Farther west, in North Dakota, the annual rainfall drops to about 18 inches, and no balsam fir is found. While the increasing dryness of the air influences the western distribution of balsam fir, the increasing temperature controls its southern distribution, limiting it to higher and higher elevations the farther south it extends, until it gives way to Frazer fir (*Abies frazeri* (Pursh.) Lindl.) on the highest mountains of West Virginia, North Carolina, and Tennessee.

---

[1] John Macoun. Geological and Natural History Survey of Canada: Catalogue of Canadian Plants. Part III—Apetalæ, p. 473.

The northern limit of balsam fir's botanical range extends from Labrador and Newfoundland southwestward, crossing James Bay at latitude 54° north and, keeping slightly south of Hudson Bay, passing between Fort Severn and Front Lake to Hayes River. From

FIG. 1.—Distribution of balsam fir.

this point it turns abruptly again southward and crosses Nelson River at the outlet of Sipiwesk Lake; thence it takes a northwesterly direction to the Great Bear Lake region until it reaches and probably crosses Mackenzie River. The most northern point at which balsam fir has been observed is 62°.

Southward balsam fir is found almost all over Canada, particularly in its maritime provinces—Quebec and Ontario—in northern New England, and in the northern parts of New York, Michigan, Wisconsin, Minnesota, and northeastern Iowa. Along the Appalachian Mountains it extends through western Massachusetts, over the Catskills of New York, and through western Pennsylvania to the mountains of southwestern Virginia.

The heaviest commercial stands of balsam fir are found in Canada, in Quebec and Ontario. On the Cape Breton Islands, according to Dr. Fernow,[1] balsam fir forms a solid forest, with not over 15 per cent of spruce and a small admixture of paper birch, covering a plateau of 1,000 square miles. It is estimated to compose more than 50 per cent of the forest, 150,000 square miles in extent, on the southern slope of the Laurentian shield, south of the height of land. In the United States balsam fir is found in commercial quantities in most of Maine, the northern parts of New Hampshire, Vermont, New York, and to some extent also in the swamps of northern Wisconsin, northern Michigan, and Minnesota, or, in all, over an area of approximately 35,000 square miles.

FOREST TYPES.

The same factors that control the geographical distribution of balsam fir influence to a great extent also its local occurrence. Maine, with an average summer temperature of only 62.5° F., an average winter temperature of 20° F., and a mean annual rainfall of 43 inches, presents most favorable conditions for the tree's growth, and, indeed, here balsam fir is in general more thrifty than in any other State in which it occurs. This is shown in every way—in the greater height, larger diameter, greater clear length, more cylindrical shape of the trunk, and the smoother appearance of the bark, indicating a more rapid growth.

The forest types in which balsam fir occurs in Maine, as well as throughout northern New York, New Hampshire, and Vermont, may be classified as swamp, flat, hardwood slope, and mountain top.

SWAMP.

The swamp type occupies low, poorly drained, swampy land which never becomes entirely dry, and on which sphagnum and other mosses form the predominating ground cover. In such swamps balsam fir grows in dense stands and remains exceedingly slender, but is remarkably free from injury by fungus, especially from ground rot and from wind and frost cracks. It often grows nearly pure, though commonly it is mixed with black and red spruce, white cedar, and tamarack.

On account of its small size and slow growth, the balsam fir of the swamps is of little commercial value. This slow growth may be attrib-

---

[1] Forest Problems and Forest Resources of Canada, by Dr. B. E. Fernow, University of Toronto. Proceedings of the Society of American Foresters, Vol. VII, No. 2, 1912.

uted to two causes, excess of moisture and a short growing season. The dense evergreen foliage of the coniferous trees, as well as the ground cover of moss, shields the ice which forms in the ground during winter against the rays of the sun in the spring. Thawing, and therefore the root activity of the trees, begins later in the swamps, often five weeks, than on the slopes or dry flats.

The characteristic ground cover of balsam swamps is made up of mosses, which form about 70 per cent of the herbaceous vegetation. The character of the vegetation and the relative proportion of the different species which compose the ground cover of the swamps is as follows:

Mosses (70 per cent):
Common—.
Sphagnum.
Fern moss (*Hylocomuim proliferum*).
Shaggy moss (*Hytocomuim triquitrum*).
Scale moss.
Occasional—
Crane moss (*Dicranum fuloum*).
Fern and fern allies (10 per cent):
Common— •
Spinulose shield fern (*Dryopteris spinulosa*).
Cinnamon fern (*Osmunda cinnamomea*).
Lady fern (*Asplenium felixfemina*).
Long beech fern (*Phegopteris phegopteris*).
Oak fern (*Phegopteris dryopteris*).
Marsh shield fern (*Dryopteris phegopteris*).
Crested shield fern (*Dryopteris cristata*).
Sensitive fern (*Onoclea sensibilis*).
Rare—
Fernata grape fern (*Botrychium obliquum*).
Horsetail (*Equisetum sylvaticum*).
Flowering plants (20 per cent):
Common—
Wood sorrel (*Oxalis acetosella*).
Gold thread (*Coptis trifolia*).
Bunchberry (*Cornus canadensis*).
Dalibarda (*Dalibarda repens*).

Flowering plants (20 per cent)—Continued.
Common—Continued.
Sweet white violet (*Viola blanda palustriformis*).
Creeping snowberry (*Chiogenes hispidula*).
Clintonia (*Clintonia boreatis*).
Wild sarsaparilla (*Aralia nudicaulis*).
Twin flower (*Linnaea borealis*).
Occasional—
Chickweed wintergreen (*Trientalis americana*).
Painted trillium (*Trillium undulatum*).
Two-leaved Solomon's seal (*Unifolium canadense*).
Rare—
Creeping wintergreen (*Gaultheria procumbens*).
Indian pipe (*Monotropa uniflora*).
Underbrush:
Common—
Green alder (*Alnus alnobetula*).
Mountain ash (*Pyrus americana*).
Withe rod (*Viburnum cassinoides*).
Occasional—
Mountain holly (*Ilicioides mucronata*).
Fetid currant (*Ribes prostratum*).
Swamp honeysuckle (*Lonicera oblongifolia*).
Pale laurel (*Kalmia glauca*).
Mountain maple (*Acer spicatum*).
Hobble bush (*Viburnum alnifolium*).

FLAT.

The flat type is intermediate between the swamp and the hardwood slope. It includes the low swells adjoining wet swamps, or the gentle lower ridges, and also the knolls in wet swamps. It is fairly well drained, and fern moss replaces sphagnum as the principal ground cover. In essentials it is still the swamp, except that it is drier. Lumbermen, in fact, call it "dry swamp." Here balsam grows rapidly, becomes tall, straight, and clear-boled, attains a fair diameter, and, as in the swamp, often grows pure. But the trees in the dry swamp are much more subject to ground rot than in the wet swamp. When it occurs in mixture its associates are red spruce, yellow birch, and red maple—the two latter small and unimportant. It is on the flats that the heaviest stands of balsam fir are found, and here also it grows more commonly in mixture with red spruce, with which it is cut and marketed for the same uses. Of the four types, therefore, the flat is commercially the most important.

The characteristic ground cover of the flat, in addition to leaf litter (15 per cent), is as follows:

Moss (60 per cent):
Common—
Fern moss (*Hylocomuim proliferum*).
Scale moss.
Ferns (5 per cent):
Common—
Spinulose shield fern (*Dryopteris spinulosa*).
Lady fern (*Asplenium felixfemina*).
Flowering plants (20 per cent):
Common—
Wood sorrel (*Oxalis acetosella*).
Bunchberry (*Cornus canadensis*).
Creeping wintergreen (*Chiogenes hispidula*).
Clintonia (*Clintonia borealis*).
Sarsaparilla st. (*Aralia nudicaulis*).
Dalibarda (*Dalibarda repens*).
Occasional—
Trillium (*Trillium erythrocarpum*).

Flowering plants (20 per cent)—Continued.
Occasional—Continued.
Chickweed wintergreen (*Trientalis americana*).
Rattlesnake plaintain (*Epipactis repens*).
Gold thread (*Coptis trifolia*).
Rare—
Lady's slipper (*Cypripedium acaule*).
Underbrush:
Common—
Hobble bush (*Viburnum lantanoides*).
Withe rod (*Viburnum cassinoides*).
Mountain ash (*Pyrus americana*).
Occasional—
Swamp honeysuckle (*Lonicera oblongifolia*).
Mountain maple (*Acer spicatum*).
Service berry (*Amelanchier canadensis*).
Beaked hazelnut (*Corylus rostrata*).

HARDWOOD SLOPE.

This is the best-drained type. In it hardwood leaf litter, instead of mosses, forms the chief ground cover.

On the slopes balsam fir never occurs in pure stands, but grows scatteringly among red spruce and large-sized hardwoods. The principal species of hardwoods are yellow birch, red maple, sugar maple, and beech. Here balsam fir, provided it is not too heavily shaded, grows rapidly and becomes comparatively large and tall, reaching on the slopes, in fact, its best individual development. It is apt to be very defective, however, and is especially liable to ground rot unless it grows near a brook or spring which furnishes a plentiful supply of water to its roots.

The characteristic ground cover of the hardwood slope besides leaf litter (40 per cent) is as follows:

Mosses (5 per cent):
Occasional—
Plume moss (*Hypnum crista-castrensis*).
Crane moss (*Dicranum fuloum*).
Shaggy moss (*Hylocomium triquitrum*).
Mountain fern moss (*Hylocomium proliferium*).
Fern and fern allies (30 per cent):
Common—
Spinulose shield fern (*Dryopteris spinulosa*).
Shining club moss (*Lycopodium lucidulum*).
Occasional—
Hayscented fern (*Dicksonia pilosiuscula*).
Lady fern (*Asplenium felixfemina*).
Ground pine (*Lycopodium complanatum*).
New York fern (*Aspidium noveboracense*).
Silvery spleen wort (*Asplenium thelyteroides*).
Rare—
Common polippod (*Polypodium vulgare*).
Long beech fern (*Phegopteris polypodioides*).
Flowering plants (25 per cent):
Common—
Wood sorrel (*Oxalis acetosella*).
Bunchberry (*Cornus canadensis*).
Wild sarsaparilla (*Aralia nudicaulis*).
Clintonia (*Clintonia borealis*).
Painted trillium (*Trillium erythrocarpum*).

Fowering plants (25 per cent)—Continued.
Occasional—
Two-leaved Solomon's seal (*Unifolium canadense*).
Sweet white violet (*Viola blanda palustriformis*).
Twisted stalk (*Streptopus amplexifolius*).
Indian cucumber root (*Medeola virginiana*).
Dalibarda (*Dalibarda repens*).
Gold thread (*Coptis trifolia*).
Rare—
Creeping snowberry (*Chiogenes hispidula*).
Indian pipe (*Monotropa uniflora*).
Rattlesnake plantain (*Epipactus repens*).
Lady's slipper (*Cypripedium acaule*).
Habenaria (*Habenaria*).
Underbrush:
Common—
Hobble bush (*Viburnum lantanoides*).
Mountain maple (*Acer spicatum*).
Striped maple (*Acer pennsylvanicum*).
Occasional—
Beaked hazelnut (*Corylus rostrata*).
Swamp honeysuckle (*Lonicera oblongifolia*).
Service berry (*Amelanchier canadensis*).

MOUNTAIN TOP.

Higher up the slopes, as the number of sugar maples gradually increases, balsam fir becomes more and more scattering, until it is found only as single specimens here and there, and on the middle slope, the driest portion of the mountain, disappears entirely. Approaching the top, however, at 2,500 or 3,000 feet above sea level, balsam fir reappears, often forming pure stands. Together with black spruce, it is the last to give way to the Alpine flora on mountains rising above timber line.

Conditions on a mountain top, where the prevailing low temperature retards evaporation and helps the condensation of moisture in the air, are similar to those in the swamp, and balsam fir shows much the same development in both places. The chief difference is that on the mountain top the trees are shorter. The principal ground cover is the same sphagnum moss found in the swamps. Balsam fir of the mountain top has no commercial value, because of the difficulty of lumbering it, coupled with its small size and slow growth.

Approaching timber line, balsam fir becomes dwarfed, procumbent, or spreading, with a short trunk and long, horizontal branches spreading near the ground. On the lower surfaces of the lower branches touching the ground, roots are often formed. When such a branch becomes detached from the main stock it may even give rise to an independent tree. The capacity to transform branches into roots has also been observed in balsam fir seedlings that have germinated in wet moss. Often in such cases, as the tree grows larger, additional roots are formed at the lower nodes of the stem beneath the moss, where originally branches grew.

In Michigan, Wisconsin, and Minnesota balsam fir, when growing in mixture with tamarack, arborvitæ or white cedar, spruce, aspen, or black ash, under conditions similar to those existing in the swamps of the northeastern States, is of poor development, with a diameter seldom larger than 11 inches and a height of 30 or 35 feet.

PRESENT STAND AND CUT.

The total stand of balsam fir throughout its range of commercial occurrence may be placed somewhere in the neighborhood of 5,000,000,000 board feet.

TABLE 1.—*Present stand of balsam fir, by States, in million board feet:*

Maine	3,000
New York	250
New Hampshire	400
Wisconsin	395
Michigan	200
Vermont	110
Minnesota	1,000
Total	5,355

This estimate is undoubtedly very crude, but even a crude estimate seems better than none.

Only within the last four or five years have any records been kept of the cut of balsam fir for various purposes. Growing with spruce, and being used for the same purposes, it always went under the name of spruce.

According to the census reports for 1909, the total lumber cut of balsam fir for the United States for 1909[1] amounted to 108,702,000 feet, and according to the census report for 1910, 132,362 cords, or 66,181,000 board feet, for pulp. The total annual cut of balsam fir in the United States at present is about 175,000,000 board feet. At this rate, the present stand, not counting the increment, will last for about 30 years.

MAINE.

In Maine, balsam fir is most common in the eastern part of the State, especially in the big flat country at the head of the St. John and Penobscot Rivers and their tributaries, and along the coast for about 10 miles inland, where it constitutes nearly one-fifth of the coniferous forests. In the western part of the State, along the Androscoggin and Kennebec Rivers, its proportion in the forest is comparatively small.

From actual measurements by the Forest Service, extended over many hundred acres and upon estimates obtained from persons most familiar with the Maine forests, it is safe to assume that balsam fir constitutes in volume for the whole State not less than 15 per cent of the spruce stand. Based upon an estimate by the Maine forest commissioner in his annual report for 1902, which gives the present stand of spruce as 21,239,000,000 feet, the present stand of balsam fir in Maine approximates 3,000,000,000 board feet.

Replies to circular letters sent out in 1903 by the Forest Service to all saw and pulp mills in Maine, regarding the use of balsam fir, justify the conclusion that about 70,000,000 board feet of this species is being cut annually for pulp and lumber. This estimate is confirmed by the statistics of the Bureau of the Census, which show that in 1910, 32,861 cords, or approximately 16,500,000 board feet,[2] of balsam fir was cut for pulp in Maine, and that in 1909 nearly 50,500,000 board feet was cut for lumber. This would make the total annual cut of balsam fir in Maine about 67,000,000 board feet. The amount of balsam fir used by the sawmills appears to be proportionately larger than the amount used by the pulp mills. This is undoubtedly due to the great amount of spruce used for pulp. Pulp

---

[1] The total cut of balsam fir for lumber in 1910 was 74,580,000 board feet, but this figure does not include the cut in the State of New York, and therefore is incomplete For this reason the figures for 1909 were used.

[2] In converting cords into board feet, 2 cords are taken to be equal to 1,000 board feet.

manufacturers can afford to pay stumpage prices for spruce which places it almost beyond the reach of the lumbermen. The latter, therefore, must turn more and more to other species, such as hemlock and balsam fir, at least for those purposes for which they will serve as well as spruce.

The amount of balsam fir used by the sawmills has increased within the last 10 years more than 50 per cent, and in some places even 75 or 100 per cent. Ten or 15 years ago, in fact, hardly any balsam fir not large enough for saw logs was cut; now it is taken almost as readily as spruce.

NEW YORK.

In northern New York, balsam fir is abundant in Franklin, Warren, Oneida, Lewis, and Clinton Counties, though it is not lacking in any township throughout the whole Adirondack region. It constitutes at present about 7 per cent of the "spruce" product and about 10 per cent of all the "spruce" pulpwood cut in the Adirondacks. Since balsam fir is now cut for pulp as readily as spruce, and practically no discrimination is made between the two, its proportion in the total output of pulpwood serves to indicate its proportion in the standing coniferous timber. Actual measurements over many acres in different parts of the mountains confirm this representation of balsam fir in the Adirondack forest. A distinction must be made, however, between the numerical and the volume representation of balsam fir. Numerically balsam fir constitutes from 20 to 50 per cent of the total stand, yet, since it never reaches the same sizes as spruce, its proportion by volume must necessarily be less. Based upon figures of the United States Census for 1900 on the stand of coniferous timber in the Adirondacks, the present stand of balsam fir in the Adirondack forests must be between 250,000,000 and 300,000,000 board feet.

The cut of balsam fir in the Adirondacks in 1910 amounted to 33,504,500 board feet, of which 9,248,000[1] board feet were cut for lumber and 24,256,500 board feet (48,513 cords) for pulp. The greater use of balsam fir by the pulp manufacturers than by the sawmills in the Adirondacks is explained by the leading place which New York State occupies in the pulp industry and by the decreased supplies of spruce, necessitating the use of all coniferous timber available for pulpwood.

NEW HAMPSHIRE.

In New Hampshire balsam fir is found mainly in the northern part of the State—in the White Mountains and in upper Coos County. In the southern part of the State it is found in any quantity only in the large swamps around the sources of the Contoocook and Millers

---

[1] This figure is for 1909; as no figures are available regarding the balsam fir cut for lumber in 1910, it is used as the nearest figure available.

Rivers. Altogether it constitutes about 10 per cent of the total so-called spruce cut for pulpwood in northern New Hampshire and from 1 to 5 per cent in the rest of the State. Since 97.4 per cent of the total cut of pulpwood in New Hampshire comes from the northern portion, 9 per cent may be considered a fair average proportion of balsam fir in the total output of pulpwood in the State.

The percentage of balsam fir used in mixture with spruce in the sawmills varies, according to the location of the mill, from 1 to 20, being largest in Coos County; but for the whole State it probably does not exceed 5 per cent. Thus, about 5,700,000 board feet of balsam fir were cut for pulpwood (1910) and about 12,200,000 board feet for lumber (1909), making a total of 17,900,000 board feet.

Accepting the present stand of softwoods in the four main drainage systems of northern New Hampshire as in the neighborhood of 4,764,000,000 board feet, the present stand of balsam fir in New Hampshire may be estimated in round figures to be 400,000,000 board feet.[1]

VERMONT.

In Vermont balsam fir is most common in the northern counties, Caledonia, Essex, and Orange containing nearly 20 per cent of the coniferous forests. In the southern half of the State balsam fir is found in any quantity only in the mountain townships. In 1910 balsam fir made up about $8\frac{1}{2}$ per cent of the total cut for pulpwood and lumber in the State. Assuming that it forms only 7 per cent of the spruce forest, the present stand of balsam fir, based on the census figures for the spruce stand in 1900, must be about 110,000,000 board feet. The annual cut of balsam fir, according to the census report for 1910, is about 12,000,000 board feet, of which about 4,000,000 board feet is for pulpwood and 8,000,000 for lumber.

WISCONSIN.

The only estimate of balsam fir in Wisconsin is that of Filibert Roth,[2] who placed the total stand in 1897 at 395,000,000 board feet (790,000 cords). In this estimate was included everything from 4 inches up. The yield per acre in all forests where balsam fir occurred was placed at from 50 to 100 board feet, or 4 to 8 cords, per 40 acres, an estimate which agreed with one made by the Chicago & North Western Railway Co. in Forest and adjoining counties. Balsam fir is thinly scattered in most forests of Wisconsin on the more humid loam and clay lands. It is generally less than 12 inches in diameter and below 60 feet in height. Table 2 gives estimates of the stand of balsam fir in the different counties in which it grows.

---

[1] Forest Service Bulletin 55, Forest Conditions in Northern New Hampshire.

[2] Forestry Conditions and Interests of Wisconsin, by Filibert Roth. Bulletin 16, U. S. Department of Agriculture, Division of Forestry, 1898.

TABLE 2.—*Stand of balsam fir in Wisconsin, by counties, in million feet board measure.*

Ashland	20	Oconto	15
Bayfield	25	Oneida	10
Chippewa	20	Portage	5
Clark	5	Price	15
Douglas	30	Sawyer	25
Florence	15	Shawano	20
Forest	40	Taylor	30
Iron	15	Vilas	10
Langlade	30	Wood	5
Lincoln	25		
Marathon	25	Total	395
Marinette	10		

The cut of balsam fir in Wisconsin for both lumber and pulp is increasing. In 1910, 4,196,000 board feet were cut for lumber and 8,502,000 board feet for pulp, a total of 12,698,000 board feet.

MINNESOTA.

In Minnesota balsam fir is confined largely to the northeastern half of the State, extending south as far as Isanti and Chisago Counties. On moist, retentive soils it reaches a fair development. In the northern counties it attains an average diameter breast high of 10 to 11 inches and an average volume of 51 board feet. Prof. Roth roughly estimated its stand in 1897 as 1,000,000,000 feet. While no cut is indicated for pulp, 10,147,000 board feet were cut in 1910 for lumber.

MICHIGAN.

Balsam fir occurs in the Upper Peninsula of Michigan in mixture with spruce, but there is little prospect of future supply from either species, since they occur scatteringly. Prof. Roth estimated the stand of balsam fir in 1897 at 400,000 cords, or 200,000,000 board feet. The estimates given by Prof. Roth 15 years ago of the stand of balsam fir in the States of Wisconsin, Minnesota, and Michigan were considered by him at that time too low, so their applicability to the present stand in Wisconsin, Minnesota, or Michigan may therefore be justified.

The cut in Michigan is close to that in Wisconsin and Minnesota, amounting to 10,712,000 board feet in 1910; of this, 5,925,000 board feet were cut for pulp and 4,787,000 board feet for lumber.

ECONOMIC IMPORTANCE.

BALSAM FIR PULPWOOD.

Balsam fir finds its greatest economic importance as a pulpwood. There is a close connection between the extent of the available supplies of spruce in a State and the amount of balsam fir used in the manufacture of pulp and paper. As long as there is a plentiful supply of the former, the use of balsam fir is naturally restricted.

The Forest Service, in 1903, sent a circular letter inclosing a series of questions to the pulp and paper manufacturers, lumbermen, town supervisors, and surveyors in States in which balsam fir occurs. Nearly 100 answers were received from pulp and paper mills, which throw much light upon the place of balsam fir in the economy of paper making. About 70 per cent of all the mills that reported use balsam fir in quantities varying from 2 to over 30 per cent of all the pulpwood consumed. The reasons given by those who do not use it are either that they can not get it, or that they do not like to use it "if they can detect it," or that they use some other species exclusively. The amount of balsam fir used by each mill varies from year to year, nor can it always be accurately ascertained at the mill. Spruce and balsam are invariably kept together, and the latter, after it has been barked and kept in water for any length of time, can not be readily distinguished. In general, it can be said that a greater percentage of balsam fir is used by the mills of New York (48,513 cords) than by those of Maine (32,861 cords). This is due partly to the ranking position occupied by the State of New York in the pulp industry and its relatively large number of sulphite mills capable of using an unlimited amount of balsam fir and partly also to the comparatively large supplies of spruce in Maine.

OBJECTIONS TO THE USE OF BALSAM-FIR FIBER.

The principal objection to the use of large amounts of balsam fir in the ground-pulp process is said to be on account of the pitch that covers the felts and cylinder faces. It is admitted by nearly all pulp and paper men that from 10 to 25 per cent of balsam can be used in ground pulp without lowering the grade of the paper produced. A few go even so far as to claim that a larger admixture of balsam fir—from 20 to 25 per cent—is of advantage, in that it makes the pulp "free"; that is, separates the spruce fibers during the manufacturing process and in this way allows the water to be easily drawn from the sheet. Still others claim that a satisfactory ground wood pulp can be made almost entirely of balsam. In chemical pulp, because of the acids dissolving the pitch, any amount of balsam can be used, though some claim that paper made of pulp containing a large admixture of balsam lacks strength, snap, and character. The pitch gives most trouble in freshly cut balsam, while in wood soaked in water over a season the amount is so small that it need not be taken into account. Some of the larger mills claim that after balsam fir has remained in the pond for one year any amount of it can be used.

RESIN CONTENTS.

The complaints against the larger amount of pitch in balsam fir are somewhat strange in view of the fact that the actual resin content of balsam fir is less than that of spruce. Resin in coniferous wood occurs normally in cells, of which the wood is built up as a house is built of bricks, and in the spaces between the cells, known as resin ducts, running vertically and horizontally through the wood. These resin ducts may be seen on cross sections of freshly cut wood as whiter or darker spots marked by exuded droplets of resin. On radial and tangential sections the ducts appear as fine lines or dots of different color. The difference in resin content of the different genera and species of the conifers depends mainly upon the number and size of their resin ducts. Balsam fir is one of the few conifers that lack resin ducts entirely, a thing which serves readily to distinguish it from the spruces and pines. Resin is found in the wood of balsam fir only in the interior of the cells, where it occurs in the form of small droplets. The bark of balsam fir is very rich in resin, but after the former is rossed off the wood should be freer of resin than spruce, which contains resin ducts and resin cells. Therefore the pitch, which according to all reports is the greatest drawback to balsam pulpwood, must either come from the resin in bark left on the surface of the block or else is formed in the process of grinding, in which case it is not of a resinous nature. In either event, the presence of pitch is apparently not due to any property of the wood itself.

A chemical determination of the resin contents of six spruce and of four balsam-fir sections made by the Bureau of Chemistry, United States Department of Agriculture, in 1904, gave the following results:

TABLE 3.—*Resin contents of spruce and balsam fir.*

SPRUCE.

	Moisture.	Non-volatile resins.	Volatile resins.	Total amount of resins.
	Per cent.	Per cent.	Per cent.	Per cent.
Top	5.60	0.88	0.23	1.11
Middle section	5.66	.92	.67	1.59
Butt section	6.39	.76	.49	1.25
Top	5.85	1.36	.27	1.63
Middle section	5.57	2.33	.50	2.83
Butt section	5.62	1.48	.34	1.82
Total				10.23
Average				1.70

BALSAM FIR.

	Moisture.	Non-volatile resins.	Volatile resins.	Total amount of resins.
Butt section	5.31	1.23	0.19	1.42
Middle section	5.06	.58	.15	.73
Butt section	5.01	.77	.19	.96
Middle section	4.80	.67	.48	1.15
Total				4.26
Average				1.06

Though the inferior quality of wood pulp containing a large amount of balsam fir can not be denied, it is probably not altogether due to the inferiority of the balsam wood, but to deficient knowledge of how to properly manufacture it into paper.

WORKING UP BALSAM FIBER.

There is no doubt that the fiber of balsam fir is weaker, shorter, and softer than spruce fiber; therefore the prevailing practice of working up balsam fir with spruce in both mechanical and chemical processes ordinarily results in an inferior grade of pulp, if the admixture of balsam is considerable. This is not so perceptible in the sulphite process as in the ground pulp. The wood of balsam fir, being softer, cuts more easily than spruce wood; therefore a stone of a sharpness and at a given pressure to produce good strong pulp from spruce makes poor pulp from balsam fir. With dull stones and light pressure a better quality of pulp could probably be made from balsam. Similarly, in the case of chemical pulp better results could most likely be obtained if weaker acids more suitable to the softer nature of balsam-fir fibers were used. The different properties of wood of spruce and that of balsam fir naturally suggest a different treatment of their fibers, which could best be accomplished by handling them separately. Experiments in this direction would probably open a much larger field for the use of balsam pulpwood than it now has.

SMALL YIELD OF WOOD FIBER.

Another drawback to balsam as compared with spruce is its smaller yield in pulp and paper per cord of wood. Being lighter than spruce when seasoned, it contains less wood substance per cord and so yields a smaller amount of pulp. The following figures regarding the yield of chemical and mechanical pulp per cord of spruce and balsam are based on actual experience and may be considered as average:

	Ground pulp.	Chemical pulp (sulphite).
	Pounds per cord.	*Pounds per cord.*
Spruce	1,800	1,200
Balsam fir	1,500	1,000

This drawback, however, would not exist if the stumpage price of balsam pulpwood were proportionately lower than instead of being nearly the same as that of spruce. Some mill men even claim that the only objection they have against balsam fir is its smaller yield in pulp, which, at the same stumpage price as spruce, makes its use unprofitable and discourages any attempts to improve methods of utilizing or manufacturing it.

UNSOUNDNESS.

In comparison with spruce, balsam is a short-lived tree, and is apt to become defective by the time it reaches large size. A log from a large tree which may seem apparently sound will, when cut up into blocks, often show heart rot in some portion of its length, or, still more frequently, the fibers at the center will be of soft texture, making its use uneconomical. Decayed heart is not so common in young, small-size trees, and since small logs contain more sap and produce better fiber than large ones, balsam of small diameters is not only suitable for pulpwood, but is to be preferred to the large sticks.

Knots, though more numerous in small sticks than large ones, are not a serious objection. They can be cheaply removed by passing the chipped wood through a tank of water, in which the knots sink and the wood is carried off from the surface.

Balsam fir cut in winter produces firmer and harder paper than when cut in summer.

The general tenor of nearly all the answers to the circular letter was that balsam fir is undoubtedly inferior to spruce in every respect, but that it has come into the pulp industry to stay. It fills a place in the economy of paper making, and its drawbacks are of such a nature that they may be to a great extent, if not entirely, overcome by intelligent effort.

BALSAM FIR LUMBER.

The increased demand for spruce by pulp men, who were able to pay higher prices for it than the lumbermen, compelled the latter to turn their attention to hemlock and balsam. Hemlock enters now more and more into building operations, supplanting spruce; while balsam fir, not being as strong as spruce, is relegated to uses for which strength is not a prime requirement. The total cut of balsam fir for lumber in 1909 was reported as 108,702,000 board feet.

Balsam fir is softer and more brittle than spruce; it decays rapidly in the ground, and when green does not hold nails well; but being light and tasteless it makes a very desirable box material, especially for foodstuffs. It is extensively used for cheese-box headings, staves for fish and sugar barrels, sardine cases, butter boxes, and the like. It is easily worked, and is well adapted for molding, novelty, bevel, and drop siding. It is of straighter grain than spruce, and in seasoning is less subject to warping and twisting, which makes it the better of the two woods for fence boards, small joists, planing, scantling, laths, and shingles. Its white color often makes it desirable for house finishing, and some consider it superior to spruce for violins. It saws easier, dries quicker, and is claimed to hold paint better than spruce. It has also been found to be suitable for rough lumber, flooring, ceiling, studding, crating, furniture, sheathing, children's carriages, toys, small frames, matches, square timber, excelsior, etc. In the form of box boards it yields about 10 per cent of material more to the cord than does spruce.

In 59 out of 141 sawmills which answered the circular letter, the use of balsam fir in the past few years has not perceptibly increased. Thirty-four mills now use from 10 to 40 per cent more than formerly, 30 mills from 40 to 75 per cent more, 13 mills from 75 to 100 per cent, while 2 mills use four times as much as they used three or four years ago. Only three mills report that the amount of balsam used by them has decreased.

LUMBERING BALSAM FIR.

In the Adirondacks, as well as in Maine, New Hampshire, and Vermont, the methods of cutting balsam and spruce for pulpwood differ somewhat from those used in getting out saw logs. Pulpwood is cut largely in summer and autumn, and is usually limited to a diameter of 8 inches on the stump and to 4 inches in the top. The trees are sawed close to the ground, the stump height seldom being over 1 foot. The logs are usually cut in lengths of 4 feet.

ADVANTAGES OF CUTTING INTO 4-FOOT LENGTHS.

Cutting into 4-foot lengths, when the drive is short and the stream shallow, has decided advantages over cutting long logs. The short sticks dry better, and for this reason few are lost through sinkage during the drive—a loss more common with balsam than with spruce. Green balsam logs do not float readily, and on a long drive may become water-logged and sink. Balsam logs, apparently sound at both ends, often contain rot in the center, and by having them cut into short lengths the buyer of pulpwood guards himself against defects. The owner of the forest, too, gains by cutting into short lengths, since it allows a fuller utilization of each individual tree. Thus, if the merchantable length of a tree that can be used for pulp is 22 feet, and the logs are cut into 12, 14, and 16 foot lengths, the most that could be used in such a case is a 16-foot log, leaving the remaining 6 feet to waste. On the other hand, by cutting into 4-foot lengths, two-thirds of the 6 feet would be turned into useful material. On a large cut this sort of waste may be considerable. It is true the short logs in the water will not support a man's weight, and so in many places are harder to drive, but since they seldom form jams and a smaller volume of water is needed to float them, the cost of driving 4-foot sticks for short distances is less than the cost of driving long logs. In one particular case, by changing the log lengths from 12 feet to 4 feet, the cost of driving over the same distance has been reduced from 44 cents to 10 cents per cord, besides lessening the loss through sinkage and undetected defects.

DIFFICULTIES IN LOGGING.

Compared with spruce, balsam fir is difficult and expensive to log. It is small, and therefore a gang working in a pure stand of balsam can not cut in a day as much as when working in spruce. When green it is heavier than spruce and therefore harder to snake out and handle, especially in summer in the swamps. It yields a greater per cent of cull, and in many cases the presence of rot can not be detected until the tree has been felled and cut into. It floats heavily, and many logs become water-soaked and sink, making the driving very difficult. To offset these disadvantages, and to make the use of balsam more profitable, its stumpage price should always be lower than that of spruce.

STUMPAGE PRICE AND LOGGING COSTS.

NEW YORK.

The ruling price in the Adirondacks for cutting and skidding pulpwood (long logs) is about $1.50 per cord. In this price the cutting of roads is included. The extra cost of resawing the long logs into 4-foot lengths and piling them along the log road is ordinarily 40 cents per cord, and requires, in addition to the regular crew of six men, two sawyers on the skidway. The logs, which in such cases are cut into lengths that are multiples of 4—as 12, 16, and 20—are snaked to the skidway, where they are sawed into 4-foot sticks and piled. A gang of eight men will cut, resaw, and pile from 9 to 12 cords per day. In cutting 14-foot lengths a gang of six men will cut and skid from 14 to 16 cords a day. The price of hauling varies with the distance. For two or three trip hauls per day, with 2 to 3 cords per sled, the charge is ordinarily $1.60 per cord. If the distance is short and several trips are possible the price is less. The stumpage price is a very variable quantity, ranging all the way from $2 to $3.50 per cord. Such pulpwood is supposed to contain, besides spruce, 10 per cent of balsam and 10 per cent of hemlock. As a rule, however, the percentage of balsam runs much higher. Since balsam pulpwood is hardly ever bought by itself, the price could not be determined, but it is probable that pure balsam pulpwood would command from 50 cents to $1 per cord less than the ordinary pulpwood now offered on the market. The average cost of driving can hardly be ascertained, being dependent upon the kind of stream, distance, number of logs, etc.

MAINE.

In Maine balsam fir is taken for pulp along with spruce, the only requirements being sufficient size and soundness. The scaler culls balsam closer than spruce. While a good deal of pulpwood is cut in winter and sawed into 4-foot sticks, which are piled and later hauled to water or rail on sleds, there is generally no difference in the methods of logging for pulp or lumber, except, perhaps, that the former is marked by closer utilization. The trees are usually cut down and topped off with the ax. Stumps run from 1.5 to 2 feet in height; most are cut pretty close to the root swelling. Logs may be even lengths up to 40 or 50 feet. In a pulp cut, however, the lengths are not carefully measured.

The stumpage price of balsam when not cut with spruce is in the neighborhood of $3.50 per 1,000 board feet, while spruce stumpage ranges from $4 to $7, a conservative average being about $5. Timber more than one-half mile from a landing is yarded; that is, put in piles of 20,000 to 50,000 board feet, and is hauled in February and March, when the snow is good. Hauling costs 50 cents per 1,000 board feet per mile. In addition, it takes four men at the yard to shovel snow off the piles and help load. Three men are required at the landing to mark and roll the logs. Each logger within one-half mile of a landing hauls as many logs as possible direct to the landing without yarding; this saves the cost of handling the logs twice. Thus, while the cost of hauling direct to the landing may not be over $4 per 1,000 board feet, yarding and then hauling increases the cost of getting out the logs to the landing to about $7 per 1,000 board feet. This cost, however, varies with the number and size of the logs, the distance to drag or haul, and the ease with which the timber can be reached. Dense undergrowth, necessitating the addition of one or more swampers to the crew, will, for instance, increase the cost of getting logs to the landing.

From $6.50 to $7 ought to cover, on an average, the cost of getting logs to the landing. Long drives, interrupted by large stretches of dead water, make driving an important item in Maine. There are two kinds of log drives, brook and river. In a brook drive the logs are driven by the individual lumberman; river driving is done by a corporation composed of the lumbermen who have logs in the river.

Balsam is driven along with spruce and, except for its greater sinkage on long drives, behaves in almost the same way. It seldom causes a jam, for if a balsam log gets crosswise in a bad place it usually breaks. Spruce, on the other hand, would hang and perhaps start a jam.

NEW HAMPSHIRE AND VERMONT.

In New Hampshire and Vermont methods of logging essentially resemble those of Maine, but in places acquire some of the New York features of pulpwood cutting. Occasionally both are modified to meet local conditions.

In order to ascertain roughly the weight of a cord of green and dry balsam and spruce pulpwood, five balsam firs and five spruces were felled, and three sections, each equal to a quarter of a cubic foot, were taken from the bottom, base of the crown, and top of each tree, and their weights determined at the time of cutting, and again two weeks and three weeks later. From these weights the average weight of 1 cubic foot of green and half-seasoned spruce and balsam wood was obtained. At the same time balsam and spruce were piled separately, and the actual cubic contents of solid wood in a cord determined. By multiplying the average weight of 1 cubic foot of green and half-seasoned balsam and spruce by the number of cubic feet of solid wood in a cord the weight of 1 cord of green and half-seasoned balsam and spruce pulpwood was obtained. From figures for weight per cubic foot given by Prof. C. S. Sargent, the weight of 1 cord of air-dry balsam and spruce was determined, respectively, as 2,252 and 2,662 pounds. The results of the different weighings are presented in Table 4.

TABLE 4.—*Weight per cubic foot of spruce and balsam fir.*

No. of tree.	Green (Sept. 5).		Half seasoned (Sept. 26).	
	Spruce.	Balsam.	Spruce.	Balsam.
	Pounds.	Pounds.	Pounds.	Pounds.
I	49.00	52.00	35.25	37.06
II	50.75	52.25	30.75	36.25
III	44.75	55.00	30.50	37.75
IV	51.00	51.25	35.00	34.00
V	44.25	46.00	32.00	32.00
Average weight per cubic foot	48.15	51.30	32.70	35.41
Average weight per cord	4,543.00	4,858.00	3,094.00	3,354.00

Thus, balsam weighs about 7 per cent more than spruce when green and 18 per cent less when dry. The sections taken from the butts of the trees weighed the least; the sections from the tops were the heaviest, due undoubtedly to the proportionately greater amount of sap and larger number of knots in the tops. Pulpwood never becomes entirely dry in the woods, and though by the time balsam is drawn to the river it loses about 30 per cent of its weight, it is still probably from 5 to 6 per cent heavier than spruce.

In the Adirondacks pulpwood is now measured almost exclusively by the cord. A cord contains 128 cubic feet of stacked wood, represented by a stack 4 feet high, 4 feet wide, and 8 feet long. In order to find the number of cords in a stack of other dimensions the length of the stack is multiplied by its width and height, and the result divided by 128. Thus, a stack 4 feet high and 8 feet long made of 12-foot logs contains 3 cords, the same as a stack 4 feet high and 24 feet long made of 4-foot sticks.

CONDITIONS AFFECTING THE SOLID CONTENTS OF WOOD IN A CORD.

LENGTH.

Though the number of cubic feet in both stacks is the same, the actual contents of solid wood is not. Logs are never entirely straight and smooth, and between them in the pile are cracks which increase in size with the length of the sticks. Thus, if 3 cords of 12-foot logs were resawed into 6-foot lengths there would not be enough wood to measure 3 cords, or a stack 4 feet high and 16 feet long. The stack would be smaller and the shrinkage even greater were the 12-foot logs resawed into 4-foot lengths. Thus, the shorter the stick the more wood is required to make a given number of cords. Careful investigation abroad showed that the difference in the solid contents of a cord made of 12-foot logs and one of 4-foot sticks amounts to at least 6 per cent. Pulpwood in the Adirondacks is cut mostly into 4, 12, and 14 foot lengths. It ought, therefore, to be of great practical interest to the owner of a forest tract, as well as to the buyer of pulpwood, whether the wood is cut and stacked into 4 or 12 foot lengths. Twenty thousand cords are frequently cut from a single tract during one year, and the choice of 4 or 12 foot lengths means a difference of 1,200 cords, or, in money (at stumpage price of $2.50 per cord), of $3,000.

DIAMETER.

The diameter of the logs also has a decided influence upon the volume of solid wood in the stack. The smaller the logs the less the amount of wood, for the more sticks in the cord the greater is the number of cracks. The difference in solid volume of two stacks, one composed of sticks twice as large as those in the other, may amount to 13 cent, and if of sticks four times as large to even 25 per cent. From 6.26 cords of pure balsam fir pulpwood, cut into 4-foot lengths, all sticks 7 inches and below in diameter at the upper end were selected and piled separately from the sticks with a diameter of more than 7 inches. To find the volume of solid wood in the two stacks the volume of each 4-foot stick was determined. The stack made of logs 7 inches and less in diameter averaged 116 sticks and 91.4 cubic feet of solid wood per cord. The stack made of logs above 7 inches in diameter averaged 56 sticks and 95.75 cubic feet, or 5 per cent, more of solid wood per cord. In another case 8.68 cords of balsam, piled and measured in the same way, gave relatively similar results.

FORM.

The smoother and straighter the logs the fewer the air spaces between them, and consequently the greater the solid contents of the stack. For this reason the clear trunks of trees yield more solid wood per given space than the tops.

SEASONING OF WOOD.

As freshly cut wood dries in the air the stack shrinks, resulting in an increase of solid wood per given space. In drying, it is true, the wood cracks, and the bark becomes detached, which tends to counteract the shrinkage of the stack, but not enough to neutralize it entirely. It therefore makes a difference how soon after felling the stack is measured. Softwood in thorough air-drying shrinks from 9 to 10 per cent, consequently stacks of dry softwood have from 9 to 10 per cent more of solid volume than similar stacks of green wood.

MANNER OF PILING.

The volume of solid wood in the stack is also affected by the way it is piled and fixed. The higher the stack, the less closely it can be piled and the less wood it will contain per given space. Stacks higher than 4 or 4.5 feet can not be piled conveniently. The heavier the log the less close is the piling and the less solid wood there is in the cord. In order to hold the pile together one or two stakes are used at each end. The volume of solid wood per cord is higher when one stake is used at each end of the stack than when two stakes are used, since in the latter case the ends of the sticks can not reach much outside the stakes. There always remains some space between the stakes and the wood, so that the fewer the stakes used for the total amount of wood corded (i. e., the longer the stacks), the higher is the solid volume per cord. Efficiency of labor, moreover, has its effect. If the branches are not trimmed close to the body of the log, if the logs are chopped instead of sawed, if the laborer is careless in piling, there is less solid wood per given space.

HOW THE STACK SHOULD BE MEASURED.

The length of a stack should be measured half way up from the ground, since the top is usually longer than the bottom, due to the spreading of the end stakes. The top length would give more and the bottom length less than the actual solid volume. The height of the stack, which is seldom uniform, should be measured at several places on both sides, and the average taken.

ACTUAL SOLID CONTENTS OF CORDS OF DIFFERENT LENGTHS AND DIAMETERS.

No correct comparison can be made, then, between stacks containing the same number of cords, but composed of logs of different lengths, diameters, or shape, unless the actual solid volume of the two stacks is known. Only by knowing this can one avoid paying the same amount of money for different amounts of solid wood. Table 5 gives the solid volume of wood in a cord according to size and length of the sticks. Other factors which influence the solid contents are variable, and are therefore not considered. Sticks with a diameter of more than 7 inches at the upper end are usually derived from the lower part of the trunk, are free from branches, and cylindrical in shape. Sticks less than 7 inches in diameter come usually from the upper parts of the trees. The mixture of these two classes is typical of most of the pulpwood offered on the market.

TABLE 5.—*Volume of solid wood per cord.*

Length.	Small diameter over 7 inches.	Small diameter from 7 to 4 inches.	First and second classes mixed.
Feet.	*Cubic feet.*	*Cubic feet.*	*Cubic feet.*
4	96.7	92.4	94.9
8	91.6	87.2	89.7
12	86.2	81.6	84.3
16	80.2	75.5	78.3

Table 5 is presented as a basis for specifications in contracts for pulpwood. Designating the money value of 1 cord of 4-foot logs of the third class as 100, the value of 1 cord of logs of the lengths and diameters given in table 6 will be as follows:

TABLE 6.—*Relative money value of cords composed of logs of different lengths and diameters.*

Length.	Small diameter over 7 inches.	Small diameter from 7 to 4 inches.	First and second classes mixed.
Feet.	*Per cent.*	*Per cent.*	*Per cent.*
4	101.8	97.4	100.0
8	96.6	91.9	94.6
12	90.9	86.0	88.9
16	84.6	79.6	82.6

LIFE HISTORY OF BALSAM FIR.

GENERAL APPEARANCE.

Balsam fir (*Abies balsamea* (Linn.) Mill.) is a small evergreen tree, seldom reaching, in the State of New York, a height of 85 feet and a diameter of 18 inches breast high. (Plate I.) In Maine occasional trees attain a height of 95 or 100 feet and a diameter of 25 or 30 inches. As a rule, however, mature trees are from 12 to 16 inches in diameter and from 70 to 80 feet high. Of all the northern softwoods, balsam fir is probably one of the most symmetrical trees. The bole has a very uniform and gradual taper closely resembling a cylinder in form.

The crown of a normal tree is always conical, since the lower branches are longer than the upper ones. The main branches are arranged in whorls of 4 to 6, with here and there scattered solitary branches between. The lower branches of a mature tree are long, slightly pendulous, those near the middle of the crown horizontal, and the upper short branches ascending. As with white pine, the branches readily die off, but remain on the trunk for a long time. The crown, therefore, may begin very high up the tree, but the clear length in the lumberman's sense is comparatively short. This explains to a large extent why balsam-fir lumber has, as a rule, more knots than spruce lumber.

FOLIAGE.

The needles differ in shape and arrangement, depending upon their position on the tree. They are sessile, narrow, linear, notched at the apex, and from half an inch in length on the upper branches to an inch and a half on the lower ones. On the lower branches, while actually spirally arranged, they are twisted so as to form but two rows, horizontally spread on each side of the branch. On the upper branches they retain their ascending spiral arrangement. They are dark green above and silvery white beneath on account of the many stomata which are arranged in lines and appear as minute, shiny dots, and are especially conspicuous in newly formed leaves. This arrangement of both branches and foliage is simply a response of the tree to light conditions. The top of a tree normally receives light from all sides, and needles and branches, therefore, stand out in all directions. At the bottom of a tree in the forest light comes mainly from above, hence the branches and needles there are arranged in a horizontal plane with their functional surface upward. Trees that are suppressed have feathery and spray-like foliage, also due to light conditions.

The foliage of balsam fir persists for from 8 to 13 years, depending upon the amount of shade and the thriftiness of the tree. Dense shade and rapid growth cause the needles to drop earlier; abundance of light and slow growth allow them to remain on the tree for a longer time.

LEAF STRUCTURE.

The leaf structure of balsam fir, as of the entire genus Abies, is very similar to that of the pines. It consists of three parts—the outer or cortical part, the chlorophyll-bearing or mesophyll part, and the fibro-vascular part. The outer part is composed of an epidermis and strengthening cells lying directly beneath. The chlorophyll part is composed of parenchyma cells, among which are distributed the resin ducts. These ducts either lie directly beneath the epidermis close to the periphery of the leaf surface or else are surrounded

by the parenchyma nearest to the center of the leaf. In the former
case the ducts are termed peripheral; in the latter, medial. The
fibro-vascular bundles lie in the center of the leaf and are surrounded
by an imperfect bundle sheath.

The leaf structure affords a reliable means for distinguishing one
species of fir from another. Only Alpine fir (*Abies lasiocarpa*) and
Fraser fir (*Abies fraseri*) are likely to be confused with balsam. The
range of balsam touches that of Alpine fir in the West and that of
Fraser fir in the South. These three species are readily distinguished

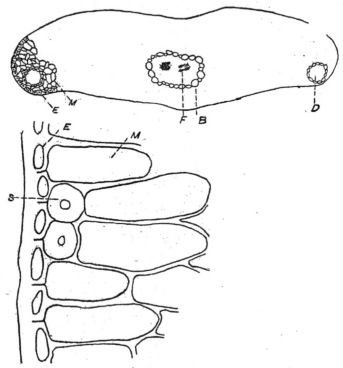

Fig. 2.—Leaf structure of *Abies grandis:* *D*, ducts; *B*, bundle sheath; *F*, fibro-vascular bundle; *M*, mesophyll; *E*, epidermis; *S*, strengthening cells.

from the rest of the firs, such as *Abies grandis* (fig. 2) and *Abies con-
color*, by the position of the resin ducts. In balsam fir (fig. 3), Alpine
fir (fig. 4), and Fraser fir they lie nearer the center, while in the other
species they lie close to the periphery of the leaf, as observed by cutting through a fir needle and observing the exudation of the resin.
Balsam is distinguished from Alpine and Fraser fir by the presence
of only a few or the entire absence of strengthening cells, which, in
the two other species, occur in considerable number.[1]

---

[1] The Resin Ducts and Strengthening Cells of Abies and Picea, by Herman B. Dorner. **Proceedings of**
Indiana Academy of Science, 1897, p. 116.

Bul. 55, U. S. Dept. of Agriculture. PLATE I.

BALSAM FIR, ADIRONDACKS, NEW YORK.

BARK.

The bark on the stump of a mature balsam fir is seldom thicker than 0.7 of an inch and in the top, at a diameter of 4 inches, seldom more than 0.3 of an inch. In volume the bark amounts to about 10.5 per cent of the whole tree. On thrifty trees it is very smooth, except for swellings or "blisters," which contain a clear liquid from which the so-called Canada balsam is obtained by distillation in water. In abundant seed years balsam blisters are very small, probably due to the tree's use of most of the foodstuffs for the production of seed. Abnormally thick, rough, or scaly bark of an ashy color, accompanied

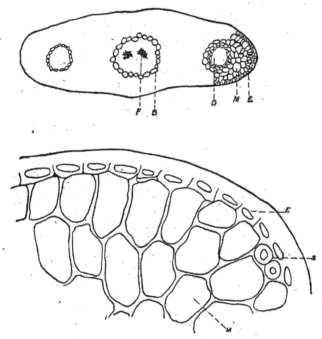

Fig. 3.—Leaf structure of *Abies balsamea:* D, ducts; B, bundle sheath; F, fibro-vascular bundle; M, mesophyll; E, epidermis; S, strengthening cells.

by swelling of the bole, is an almost infallible sign that the tree is rotten at those parts. The natural color of the bark in young trees is a dull, faded green, mottled with patches of gray. With age the bark becomes entirely gray and slightly scaled, but not the dull ashy gray of a defective tree or the shaggy moss and lichen-covered scale of a slow-growing balsam in the swamp.

ROOT SYSTEM.

Whether grown in deep or shallow soils, balsam fir produces a very superficial root system, penetrating to a depth of about 2 or 2.5 feet. Taproots, if developed at all, soon die and rot away, especially in soils lacking an abundance of moisture, and often become points of entrance for destructive ground rot. The strongly developed lateral roots extend horizontally in all directions for a distance of 4 or 5 feet, and even more. The bark of the roots is bright red and comes off in thin scales.

FLOWERS.

The female and male flowers (cones) occur on the same tree in the top at the outermost ramifications of the branches. The female flowers occupy the extreme top near the ends of the upper branches and are borne perpendicularly in the leaf axils on the upper sides of

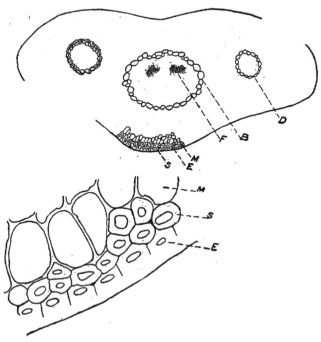

Fig. 4.—Leaf structure of *Abies lasiocarpa: D*, ducts; *B*, bundle sheath; *F*, fibro-vascular bundle; *M* mesophyll; *E*, epidermis; *S*, strengthening cells.

the previous year's branches, while the male flowers are borne mostly on the under or lower sides. The cones, which are violet in color, cylindrical shaped, and from 2 to 4 inches long, do not turn downward after fertilization, like the cones of spruce, but remain erect. They ripen in one year, about the end of September. The mere opening of the erect cones does not liberate the seeds, but the flat, smooth scales of the cone and the scale bracts themselves drop off, carrying the seed with them, and leaving the axils of the cone on the tree for years. The deciduous scales of the cone are broad, round at the top, and narrow to a wedge at the bottom. Within each scale are two winged seeds. Outside of each scale, at the bottom, is a bract [1] resembling a transformed, winged fir leaf, the end of which, on a mature cone, seldom protrudes enough to be noticed. These bracts furnish a means of distinguishing balsam, Fraser, and Alpine fir. In general, the relative lengths of the cone scale and this bract are means to distinguish between the different native firs, but in the case of balsam the value of this distinction is lessened because of the occurrence of forms with slightly exserted or protruding bracts.

The classification of Fraser fir as a distinct species rests not on the protrusion of the bract, but on its spatulate and reflexed form. The forms of balsam fir with slightly exserted bracts need not, therefore, cause any confusion, for though these do protrude a little, they are not different in shape from the included form, and are neither spatulate nor reflexed. (Fig. 5.)

Since the bracts of Alpine fir never protrude, this variant character in balsam is of value in distinguishing it from Alpine fir. Furthermore, the cone scale of Alpine fir is larger than that of balsam, as shown in Fig. 5 (b and c).

This distinction, however, can not always be relied upon, because the size and form of

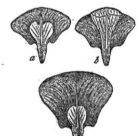

FIG. 5.—Cone scale and bract, natural size: a, Abies balsamea (L.) Mill.; b, Abies fraseri (Pursh.) Lindl.; c, Abies lasiocarpa (Hook.) Nutt.

the cone scales of Alpine fir vary. It is safer, therefore, to distinguish Alpine from balsam fir by the form of the bract, which in the former is conspicuously long pointed.

REPRODUCTION.

Under favorable conditions balsam fir bears fruit when about 20 years old and 15 feet high. Regular production of seeds, however, does not begin before the age of 30 or 35 years. On high mountains, above timber line, scrubby balsam begins to bear seeds in large quantities when from 23 to 25 years old. The amount of seeds borne by individual trees depends, of course, on the size of the crown. As a rule trees in a dense stand bear less seed than trees in the open. In a mixed forest the dominant trees are prolific seeders, the intermediate trees moderately so, while the suppressed trees produce no seed at all. Although balsam fir produces some seed every year, plentiful seed years occur only at intervals of two, three, and even four years.

---

[1] Discussion and Drawings of Cone Scales and Bracts, by William H. Lamb, Forest Service.

FREQUENCY OF SEED YEARS.

An investigation of the ages of seedlings on several reproduction plots in the Adirondacks revealed the fact that of the total number counted the per cent of ages from 1 to 11 years was as follows:

	Per cent.
1 year old (1902)	46. 1
2 years old	. 0
3 years old	. 2
4 years old	1. 2
5 years old (1898)	51. 1
6 years old	. 1
7 years old	. 1
8 years old	. 5
9 years old	. 2
10 years old	. 3
11 years old	. 2

The large representation of seedlings 1 and 5 years old serves to indicate an occurrence of plentiful seed years in the Adirondacks at intervals of four years.

Other seed years can not be readily determined by this study, since the seedlings after reaching an age of 6 years do not stand the dense shade very well, and few survive. In Maine a similar study has shown the occurrence of good seed years every other year. In one instance the seed years were traced back to 1882, all of them occurring in the even years. In New Hampshire good seed years were found to occur every third year.

QUANTITY AND QUALITY OF SEED.

As determined by the Forest Service, the number of seed per pound averages about 36,000; the weight of a thousand seeds, 0.39 ounce (12.4 grams); and the germination per cent, from 20 to 30.

GERMINATION.

Since the seeds are scattered late in the fall, when frosts have already occurred, they lie dormant through the winter and come up the next spring. Hardwood leaf litter, duff, moss, mineral soil, rotten logs—all present an equally good germinating bed, if moist. Balsam differs from spruce in this respect, requiring more moisture, as may be inferred from the fact that spruce seedlings are found in drier situations, both on logs and on the ground. A rather dry and high log will have plenty of spruce seedlings and very few balsam, while a well-rotted moist log will have a great number of balsam seedlings. The same is true of stumps.

The number of seedlings that come up on the acre varies with the type of forest. Thus on the hardwood slopes in the Adirondacks, where balsam fir occurs scatteringly, the number of seedlings per acre is small, often only 700 to 1,000, though occasionally, if there are a number of large balsams, the number may reach 50,000. The number of seedlings is, of course, largest in pure stands of balsam, where they may be 300,000 and more to the acre. In mixture with spruce in the swamps and flats the number of balsam seedlings will vary from several thousand to 200,000 and more, according to the number of large seed-bearing balsams in the stand.

TOLERANCE.

Balsam fir requires less light than tamarack, white pine, and white cedar, but more light than either red spruce or hemlock. It will, however, endure more shade on deep, moist soils than on poor, shallow ones. In mixture with spruce, mature healthy balsam invariably towers above the former. Similarly, in a mixed hardwood forest, balsam fir, when fully developed, is the dominant tree. For the first five or six years of its life, balsam will grow in dense shade, but as it develops it demands more and more light. On moist soils, however, it may thrive without being in the top story of the forest, and beneath white birch and poplar, also, it often remains apparently healthy and vigorous. But where it comes in under a hardwood forest already established, its leader is usually stunted or killed when it enters the hardwood foliage. A broken limb or leader often affords the means of entrance for rot, and though balsam, especially on deep, moist soil, is capable of recovery after a long period of suppression, it is apt in such cases to be unsound. Many trees were found to be rotten in the middle at the point of suppression, with no visible point of entrance for the rot. Others were found 100 years old, with a height of 18 feet and a diameter of 3 inches, which, after 66 years of suppression, retained sufficient vitality to grow rapidly after again receiving the light.

SOIL AND MOISTURE REQUIREMENTS.

Though their demands upon soil are very similar, balsam fir requires for its best development a richer and moister soil than does spruce. With its more northern distribution it seeks the cool and moist north and east slopes in preference to other exposures. In the Adirondacks it is hardly ever found on the abrupt, rocky, southwest slopes, with thin soil, on which spruce often forms a pure stand and reaches a good development. Balsam fir attains its best growth and largest sizes on the flats, the soil of which is usually a moderately moist, deep, sand loam. In the wet swamps with acid soils, as well as on pure sand, it thrives but poorly.

SUSCEPTIBILITY TO DISEASE AND INJURY.

FUNGI.

Balsam fir must be classed as one of the most defective of our northeastern conifers. Its chief enemies are fungi, and the weakest point of attack is the heartwood.

According to its place of origin, the rot is known as top or ground rot, and is caused by two different species of fungi, *Trametes pini* and *Polyporus schweinitzii*.[1] The latter affects the merchantable portion of the tree, and therefore does the most injury. Even when the rot does not extend far up the trunk, the tree is nevertheless lost in lumbering, since the choppers, finding the butt rotten, will in many cases leave it partly cut through, to be broken down by the first wind. Thus is the whole tree wasted, although at a short distance from the ground it might be perfectly sound. The roots are the chief points of entrance for ground rot. Ground rot is especially common in balsam on slopes in mixture with hardwoods. Its relative infrequency in the swamps is most likely due to the excess of water and poor aeration in the soil, as well as the antiseptic effect of bog water. Ground rot may also find entrance through wounds on the lateral roots. Being near the surface and extending for several feet from the base of the tree, these are readily injured in logging by falling trees or by logs dragged over them. Roots may also be wounded by sharp rocks, or they may be broken by a strong wind, or insects may puncture them. In many cases ground rot was found to be associated with deep frost cracks and holes made by ants.

Top rot, affecting the upper and less merchantable part of the tree, is less common than ground rot. It was especially noted in suppressed trees, the tops of which are often injured by rubbing against other trees, though any kind of a wound in the top may afford an entrance to the fungus. Balsam fir beneath hardwoods is often suppressed for many years, and is therefore likely to be affected by rot in the top. The same is true of dense, pure stands, in which suppressed trees eventually die from the top.

Not many opportunities were afforded to study the rate at which the rot spreads, because it was impossible to tell when a frost crack or an insect wound was made. Only wounds made by falling trees, the axe, etc., could be used. The heartwood on the stump was, as a rule, completely rotten if the wound had been made low down upon the tree from five to seven years before. During that time the rot had extended upward for from 5 to 10 feet. The rate of spreading at the top was less rapid.

---

[1] Hedgcock, George G. Notes on Some Diseases of Trees in our National Forests, II. Phytopathology 2: 77-78, April, 1912.

Since the fruiting bodies of the fungi, or, as the lumbermen call them, "punks" or "conks," appear on the fir after the tree is considerably rotten, it is exceedingly hard to tell merely by the appearance of the tree whether it is sound or not. Being short-lived, balsam fir at the age of 80 to 100 years is already old, and especially susceptible to rot of any kind. Therefore one seldom finds an old balsam that is perfectly sound.

"GLASSY" FIR.

During the winter months balsam fir logs often have on cross section a "glassy" or "icy" appearance, which some lumbermen consider an indication of defect. When cut by the crosscut saw, the wood shows irregular areas which are perfectly smooth and shiny as if planed. A microscopical examination of the wood,[1] however, did not reveal any signs of decay in the smooth areas, and the structure could not be distinguished from that of the ordinary rough areas. During winter the water present in the wood of balsam fir is mostly frozen, and the shiny, smooth spots are therefore not due to any disease, but to the frozen condition of the wood. That this is so is further shown by the fact that the same section of wood when cut in an unfrozen condition appears rough over its entire area. The ice formed in the wood acts as reenforcing material and prevents the usual tearing of the wood fiber.

FIRE.

Balsam fir is very sensitive to fire. Its superficial roots are easily affected by surface fires, and the flames reach its cambium through the thin, tender bark, killing the tree. In a balsam injured by fire the lower foliage first turns brown, and finally the top. The dying in some cases is very slow, but is none the less certain.

WIND.

Balsam fir does not suffer from windshake, but it is easily uprooted and broken by wind because of its shallow root system and slender, brittle bole.

THE WOOD.

GENERAL STRUCTURE.

The wood of the balsam fir in external appearance is strikingly like that of eastern spruce, and it is often necessary to go to the gross and minute characters of its anatomical structure in order to distinguish it. Balsam fir is ordinarily close-grained and, like

[1] Glassy Fir, by Hermann von Schrenk. Sixteenth Annual Report of the Missouri Botanical Garden, pp. 117–120. St. Louis, Mo., 1905.

spruce, has no distinct heartwood and sapwood. Its narrow pith rays of a pale or whitish color are scarcely visible. Air-dry wood of balsam fir is light, weighing 24 pounds per cubic foot, as compared with 28 pounds for spruce. When completely dry, it has an average density of 0.38, and loses about 4 per cent of its volume in seasoning.

COMPARATIVE LENGTH OF WOOD FIBERS OF BALSAM FIR AND SPRUCES.

Table 7 gives the average, maximum, and minimum lengths of the wood fibers of balsam fir and the northeastern spruces.

TABLE 7.—*Average. maximum, and minimum lengths of fibers of balsam fir and the northeastern spruces.*

Species.	Length of wood fiber (millimeters).		
	Average.	Maximum.	Minimum.
Balsam fir (*Abies balsamea*)	2.518	3.750	1.680
White spruce (*Picea canadensis*)	3.556	4.704	2.520
Red spruce (*Picea rubens*)	3.283	4.158	1.890
Black spruce (*Picea mariana*)	2.599	3.738	2.142

GROWTH.

Balsam fir is a fairly rapid growing tree, though not as rapid as tamarack and white pine.

HEIGHT GROWTH.

Balsam fir has a period of comparatively slow growth, which, under favorable light conditions, lasts only for the first five years of its life; a period of rapid growth then sets in and continues until the tree is 60 years old. From then on the growth in height begins to decline, and at 80 years the growth is practically at a standstill. At 150 years it stops altogether. The most rapid growth in height takes place between the twentieth and fortieth years.

The slow growth of balsam fir for the first five or six years is an inherent characteristic of the species, and occurs even under the best light conditions. Beneath the shade of other trees, however, the period of slow growth is often extended to 25 years or more because of the retarding effect of the shade.

TABLE 8.—*Comparative growth of balsam fir seedlings, in Franklin County, N. Y., in the shade[1] and in full light.[2]*

Age (years).	Height of average trees under average conditions of shade.	Height of best trees under full light.		Age (years).	Height of average trees under average conditions of shade.	Height of best trees under full light.
	Feet.	Feet.			Feet.	Feet.
1	0.1	0.1		10	1.6	3.6
2	.2	.2		11	1.9	4.3
3	.3	.4		12	2.1	5.1
4	.5	.7		13	2.4	5.8
5	.6	1.0		14	2.6	6.6
6	.7	1.5		15	2.8	7.4
7	.9	1.9		16	3.1	8.2
8	1.1	2.5		17	3.3	8.9
9	1.3	3.0		18	3.5	9.7

[1] Based on 324 trees.　　　　　　[2] Based on 104 trees.

Thus, with conditions of growth obtaining under forest management, the growth in height of balsam fir would be increased more than two and one-half times during the first 18 years of its life (9.7 feet as compared with 3.5 feet).

Tables 9 and 10 give the average growth in height on flat, swamp, and hardwood slope in the State of New York, based on age.

TABLE 9.—*Height growth of balsam fir in New York, on the basis of age, on flat, swamp, and hardwood slope.*

Age (years).	Flat. (Based on 248 trees.)			Swamp. (Based on 158 trees.)			Hardwood slope. (Based on 277 trees.)		
	Maximum.	Minimum.	Average.	Maximum.	Minimum.	Average.	Maximum.	Minimum.	Average.
	Feet.	Feet.	Feet.	Feet.	Feet.	Feet.	Feet.	Feet.	Feet.
10	8	3	5	10	3	4	8	3	5
15	13	7	9	15	6	7	15	6	9
20	18	11	14	19	8	11	22	9	14
25	24	15	19	24	10	15	28	13	19
30	29	19	23	28	12	19	34	17	24
35	34	22	27	32	14	23	40	21	30
40	39	25	31	36	16	27	45	25	35
45	44	27	35	39	18	30	50	29	39
50	49	29	38	42	20	32	54	32	43
55	52	30	41	45	21	34	58	35	47
60	56	32	43	48	23	36	61	38	49
65	59	33	45	51	24	38	64	40	52
70	61	34	47	53	25	40	66	42	54
75	63	35	49	55	26	41	68	43	56
80	65	36	51	57	27	43	69	45	57
85	67	37	52	59	28	44	70	45	58
90	68	38	53	60	29	46	71	46	59
95	69	39	54	62	30	47	72	47	60
100	70	39	55	63	30	48	73	47	61
105	71	40	56	64	31	49	73	48	61
110	72	41	57	65	32	50	74	48	62
115	72	42	58	66	32	51	74	49	62
120	73	42	59	67	33	52	75	49	63
125	74	43	59	68	34	52	75	49	63
130	74	44	60	69	34	53	75	50	63
135	74	45	60	70	35	53	76	50	63
140	75	46	61	70	35	54	76	50	63
145	75	46	62	71	36	54	76	51	64
150	75	47	62	72	37	55	76	51	64

TABLE 10.—*Average growth of balsam fir in New York, on the basis of age, on flat, swamp, and hardwood slope.*

Age (years).	Flat. (Based on 248 trees.)			Swamp. (Based on 155 trees.)			Hardwood slope. (Based on 277 trees.)		
	Height.	Growth in height every 5 years.	Annual growth in height within 5-year period.	Height.	Growth in height every 5 years.	Annual growth in height within 5-year period.	Height.	Growth in height every 5 years.	Annual growth in height within 5-year period.
	Feet.	Feet.	Feet.	Feet.	Feet.	Feet.	Feet.	Feet.	Feet.
10	5			4			5		
15	9	4	0.8	7	3	0.6	9	4	0.8
20	14	5	1.0	11	4	.8	14	5	1.0
25	19	5	1.0	15	4	.8	19	5	1.0
30	23	4	.8	19	4	.8	24	5	1.0
35	27	4	.8	23	4	.8	30	6	1.2
40	31	4	.8	27	4	.8	35	5	1.0
45	35	4	.8	30	3	.6	39	4	.8
50	38	3	.6	32	2	.4	43	4	.8
55	41	3	.6	34	2	.4	47	4	.8
60	43	2	.4	36	2	.4	49	2	.4
65	45	2	.4	38	2	.4	52	3	.6
70	47	2	.4	40	2	.4	54	2	.4
75	49	2	.4	41	1	.2	56	2	.4
80	51	2	.4	43	2	.4	57	1	.2
85	52	1	.2	44	1	.2	58	1	.2
90	53	1	.2	46	2	.4	59	1	.2
95	54	1	.2	47	1	.2	60	1	.2
100	55	1	.2	48	1	.2	61	1	.2
105	56	1	.2	49	1	.2	61	0	.0
110	57	1	.2	50	1	.2	62	1	.2
115	58	1	.2	51	1	.2	62	0	.0
120	59	1	.2	52	1	.2	63	1	.2
125	59	0	.0	52	0	.0	63	0	.0
130	60	1	.2	53	1	.2	63	0	.0
135	60	0	.0	53	0	.0	63	0	.0
140	61	1	.2	54	1	.2	63	0	.0
145	62	1	.2	54	0	.0	64	1	.2
150	62	0	.0	55	1	.2	64	0	.0

In New York balsam fir grows in height at an average rate for all types of 0.4 of a foot a year.

On flats the growth between the ages of 20 and 45 is nearly 1 foot a year. At 60 years the current annual growth equals the average annual growth, namely 0.4 foot, which indicates that at this age the annual growth begins to decline. At the age of 85 the current annual growth is only 0.2 of a foot, and at 125 years has practically stopped.

In the swamp the growth in general is slower and on the hardwood slope faster than on the flat, but on the whole it culminates and begins to decline at about the same age in all three types.

In Maine (Table 11) the average tree grows faster than in New York; namely, at the rate of 0.7 of a foot a year. The period of most rapid growth is longer from the twentieth to the fiftieth year and the total height is greater.

TABLE 11.—*Height growth of balsam fir in Maine, on the basis of age, based on 456 trees.*

Age (years).	Height of tree (feet).			Annual growth within 5-year period (feet).	Age (years).	Height of tree (feet).			Annual growth within 5-year period (feet).
	Maxi-mum.	Mini-mum.	Aver-age.			Maxi-mum.	Mini-mum.	Aver-age.	
15............	10	....	8	0.5	60............	69	21	52	0.7
20............	22	6	14	1.2	65............	72	24	56	.7
25............	33	7	20	1.2	70............	74	27	59	.6
30............	42	8	25	1.0	75............	76	31	61	.5
35............	49	10	30	1.0	80............	78	35	64	.5
40............	55	12	36	1.0	85............	79	38	66	.4
45............	60	14	40	1.0	90............	81	41	68	.4
50............	63	16	45	1.0	95............	82	45	70	.4
55............	67	18	49	.8	100............	83	48	71	.2

Table 12 shows the relation between the height and diameter growth for all types together in New York, Maine, New Hampshire, and Minnesota.

TABLE 12.—*Comparative height growth of balsam fir in different States, on the basis of diameter breast high.*

Diameter breast high (inches).	Height of tree (feet).			
	New York.[1]	Maine.[2]	New Hampshire.[3]	Minnesota.[4]
1..................................	9	12	15	8
2..................................	17	20	24	16
3..................................	26	27	31	23
4..................................	33	35	37	31
5..................................	40	42	42	37
6..................................	46	48	46	43
7..................................	51	54	50	48
8..................................	54	60	53	53
9..................................	58	64	56	58
10.................................	60	68	59	62
11.................................	63	72	61	67
12.................................	65	75	63	....
13.................................	67	78	65	....
14.................................	70	81	67	....
15.................................	72	84	69	....
16.................................	74	86	70	....
17.................................	76	....	72	....

[1] All types, based on 1,138 trees.
[2] All types, based on 456 trees.
[3] All types, based on 326 trees.
[4] All types, based on 165 trees.

These figures indicate again that the tree reaches its best development in Maine and its next best in New York. Growth in Minnesota, though apparently more rapid than in New Hampshire or New York, on the whole is poorer than in any other State. The actual number of trees on which the figures for Minnesota are based is not large, while the figures for height growth in New Hampshire are based not on actual measurements of felled trees but on those of standing trees by means of a height measurer. If the measurements in the two States had been taken in the same way and on the same number of trees, the difference in favor of Minnesota would have been eliminated.

DIAMETER GROWTH.

Balsam fir makes its most rapid growth in diameter between the ages of 25 and 70 years, during which time the average rate is about 0.11 of an inch a year, or 1 inch in 9 years. On less favorable situations it grows at the rate of about 1 inch in 10 years. After the seventieth year the diameter growth begins to decline, and at 75 years the current annual growth falls below the mean annual growth.

Tables 13 and 14 give the growth of balsam fir in diameter on different stations in the Adirondacks.

TABLE 13.—*Diameter growth, in inches, of balsam fir in New York, on the basis of age.*

Age (years).	Hardwood slope. (Based on 277 trees.)			Flat. (Based on 246 trees.)			Swamp. (Based on 158 trees.)		
	Diameter breast high.	Growth in diameter every 5 years.	Annual growth in diameter within 5-year period.	Diameter breast high.	Growth in diameter every 5 years.	Annual growth in diameter within 5-year period.	Diameter breast high.	Growth in diameter every 5 years.	Annual growth in diameter within 5-year period.
10	0.4	....	....	0.4	....	....	0.3	....	....
15	.8	0.4	0.08	.8	0.4	0.08	.6	0.3	0.06
20	1.3	.5	.10	1.2	.4	.08	1.0	.4	.08
25	1.9	.6	.12	1.7	.5	.10	1.4	.4	.08
30	2.6	.7	.14	2.3	.6	.12	1.8	.4	.08
35	3.3	.7	.14	2.9	.6	.12	2.3	.5	.10
40	4.0	.7	.14	3.6	.7	.14	2.8	.5	.10
45	4.7	.7	.14	4.3	.7	.14	3.3	.5	.10
50	5.4	.7	.14	4.9	.6	.12	3.8	.5	.10
55	6.0	.6	.12	5.5	.6	.12	4.4	.6	.12
60	6.5	.5	.10	6.0	.5	.10	4.9	.5	.10
65	7.0	.5	.10	6.5	.5	.10	5.4	.5	.10
70	7.4	.4	.08	6.9	.4	.08	5.8	.4	.08
75	7.8	.4	.08	7.2	.3	.06	6.1	.3	.06
80	8.2	.4	.08	7.5	.3	.06	6.5	.4	.08
85	8.5	.3	.06	7.8	.3	.06	6.8	.3	.06
90	8.9	.4	.08	8.1	.3	.06	7.1	.3	.06
95	9.2	.3	.06	8.3	.2	.04	7.4	.3	.06
100	9.5	.3	.06	8.6	.3	.06	7.6	.2	.04
105	9.8	.3	.06	8.9	.3	.06	7.9	.3	.06
110	10.1	.3	.06	9.1	.2	.04	8.1	.2	.04
115	10.4	.3	.06	9.4	.3	.06	8.4	.3	.06
120	10.7	.3	.06	9.6	.2	.04	8.6	.2	.04
125	11.0	.3	.06	9.9	.3	.06	8.8	.2	.04
130	11.3	.3	.06	10.1	.2	.04	9.1	.3	.06
135	11.6	.3	.06	10.3	.2	.04	9.3	.2	.04
140	11.9	.3	.06	10.5	.2	.04	9.5	.2	.04
145	12.2	.3	.06	10.8	.3	.06	9.8	.3	.06
150	12.5	.3	.06	11.0	.2	.04	10.0	.2	.04

TABLE 14.—*Number of years required by balsam fir in New York to grow 1 inch.*

Diameter breast high (inches).	Hardwood slope. (Based on 277 trees.)		Flat. (Based on 246 trees.)		Swamp. (Based on 158 trees.)	
	Age (years).	Years required to grow 1 inch.	Age (years).	Years required to grow 1 inch.	Age (years).	Years required to grow 1 inch.
1	17	17	18	18	20	20
2	26	9	28	10	32	12
3	33	7	36	8	42	10
4	40	7	43	7	52	10
5	47	7	51	8	61	9
6	55	8	60	9	73	12
7	65	10	72	12	89	16
8	78	13	89	17	108	19
9	93	15	108	19	129	21
10	108	15	128	20	150	21
11	125	17	150	22	....	....
12	142	17	....	....	....	....

Bul. 55, U. S. Dept. of Agriculture. PLATE II.

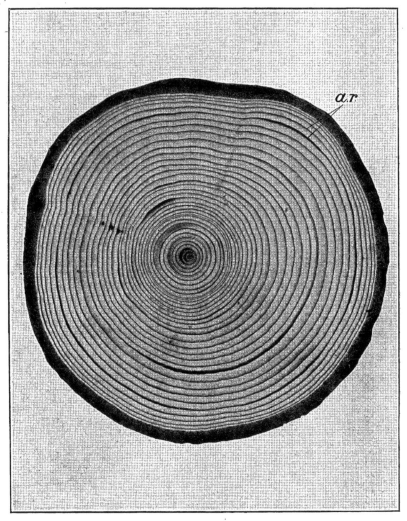

TRANSVERSE SECTION OF THE STEM OF A YOUNG BALSAM FIR TREE, SHOWING
ANNUAL RINGS OF GROWTH, *a. r.* NATURAL SIZE.

As shown by the tables, the best growth is made on hardwood slope; the poorest in swamps.

Table 15 shows the diameter growth of balsam fir in Maine for all types. The average growth here equals the best growth in the Adirondacks. On the whole, however, the periods of most rapid and slowest growth coincide.

TABLE 15.—*Diameter growth of balsam fir in Maine, on the basis of age.* _ (*All types, based on 456 trees.*)

Age (years).	Diameter breast high.	Growth in diameter every 5 years.	Annual growth in diameter within 5-year period.	Age (years).	Diameter breast high.	Growth in diameter every 5 years.	Annual growth in diameter within 5-year period.
	Inches.	Inches.	Inches.		Inches.	Inches.	Inches.
15	0.4	....	....	85	9.4	0.5	0.10
20	1.2	0.8	0.16	90	9.9	.5	.10
25	2.0	.8	.16	95	10.4	.5	.10
30	2.7	.7	.14	100	10.7	.3	.06
35	3.4	.7	.14	105	11.0	.3	.06
40	4.1	.7	.14	110	11.3	.3	.06
45	4.8	.7	.14	115	11.6	.3	.06
50	5.5	.7	.14	120	11.9	.3	.06
55	6.1	.6	.12	125	12.1	.2	.04
60	6.7	.6	.12	130	12.3	.2	.04
65	7.3	.6	.12	135	12.5	.2	.04
70	7.9	.6	.12	140	12.6	.1	.02
75	8.4	.5	.10	145	12.8	.2	.04
80	8.9	.5	.10	150	12.9	.1	.02

SUMMARY.

Diameter breast high (inches).	Age (years).	Years required to grow 1 inch.	Diameter breast high (inches).	Age (years).	Years required to grow 1 inch.
1	19	19	8	71	9
2	25	6	9	81	10
3	32	7	10	91	10
4	39	7	11	105	14
5	47	8	12	123	18
6	54	7	13	152	29
7	62	8			

EFFECT OF OPENING UP FOREST UPON DIAMETER GROWTH.

That the diameter growth of balsam fir is stimulated by opening up the forest is indicated by measurements of trees growing on uncut and on culled land (Table 16).

TABLE 16.—*Diameter growth of balsam fir, Grafton County, N. H., on uncut and culled land.*

Present diameter breast high (inches).	Diameter, in inches, breast high after 10 years.		Present diameter breast high (inches).	Diameter, in inches, breast high after 10 years.	
	Uncut land.	Culled land.		Uncut land.	Culled land.
6	6.54	6.80	10	11.18	11.40
7		7.80	11	11.88	12.34
8	9.26	9.56	12	12.88	13.84
9	10.20	10.60	13	....	14.84

COMPARATIVE GROWTH OF SPRUCE AND BALSAM FIR.

Since spruce and balsam fir nearly always grow together, and any plan of management for one species must necessarily include the other, a comparison of the growth of the two species is essential.

In Table 17 is contrasted the average growth in height and diameter of balsam fir and spruce in the State of Maine.

TABLE 17.—*Comparative growth, in height and diameter, of balsam fir and red spruce in Maine.*

GROWTH IN HEIGHT.

Diameter breast high (inches).	Height (feet).		Diameter breast high (inches).	Height (feet).	
	Red spruce. (Based on 485 trees.)	Balsam fir. (Based on 456 trees.)		Red spruce. (Based on 485 trees.)	Balsam fir. (Based on 456 trees.)
1	7	12	10	52	68
2	14	20	11	55	72
3	21	27	12	58	75
4	28	35	13	60	78
5	33	42	14	63	81
6	37	48	15	65	84
7	41	54	16	67	86
8	44	60	17	68	
9	48	64	18	70	

GROWTH IN DIAMETER.

Age (years).	Diameter breast high (inches).		Age (years).	Diameter breast high (inches).	
	Red spruce. (Based on 274 trees.)	Balsam fir. (Based on 456 trees.)		Red spruce. (Based on 274 trees.)	Balsam fir. (Based on 456 trees.)
20	0.1	1.2	90	2.7	9.9
30	.6	2.7	100	3.2	10.7
40	.8	4.1	110	3.7	11.3
50	1.1	5.5	120	4.3	11.9
60	1.5	6.7	130	4.9	12.3
70	1.8	7.9	140	5.5	12.6
80	2.2	8.9	150	6.2	12.9

Red spruce grows in height much slower than balsam fir for the first 70 years. At a diameter of about 8 inches its rate of growth in height is approximately the same as that of balsam fir. At a diameter of 12 inches balsam fir reaches almost its full height, while spruce is still far below its fullest development. From that time on spruce continues to grow at a uniform rate for a long period, while the growth of balsam fir is rapidly declining until at a diameter of about 16 inches it practically ceases.

The same is true of the growth in diameter. At the age of 100 years spruce is only 3.2 inches in diameter breast high, while balsam fir has made nearly two-thirds of its entire diameter growth. After the age of 70 years the annual growth of balsam fir declines, while that of spruce shows a gradual increase. After the age of 150 years spruce catches up with balsam fir, and finally surpasses it both in height and diameter. On the whole the rate of growth of balsam fir is more rapid during its entire life than that of spruce. The growth of spruce is, however, more persistent, and does not exhaust itself as early. It is this persistent growth and its long life which enable spruce to reach larger dimensions.

This difference in growth is also apparent on cut-over land. Measurements in New Hampshire during 10 years following cutting gave the results shown in Table 18.

TABLE 18.—*Comparative growth in diameter of spruce and balsam on culled land in Grafton County, N. H.*

Diameter breast high at time of cutting (inches).	Diameter breast high after 10 years (inches).		Diameter breast high at time of cutting (inches).	Diameter breast high after 10 years (inches).	
	Spruce.	Balsam.		Spruce.	Balsam.
8	8.82	9.56	14	14.64	
9	10.00	10.60	15	15.64	
10	11.00	11.40	16	16.64	
11	12.00	12.34	17	17.64	
12	13.00	13.34	18	18.64	
13	14.00	14.34			

Balsam fir up to 13 inches in diameter responded to increased light and space more vigorously than spruce, but did not go beyond the limit of 14 inches, while spruce continued to show a slower but a uniform increase in growth of 1 inch for each 1 inch in diameter up to 18 inches.

VOLUME GROWTH.

Tables 19 to 23 give the increment of balsam fir in cubic feet and board measure for the three different types in New York and in cubic feet for all types in Maine. The tables of volume growth, more than the tables of height and diameter growth, bring out the better development of balsam fir in Maine than in New York and other States. The annual increment in Maine is practically twice that in New York. Similarly, the volume-growth tables bring out more clearly the differences in the increment of balsam fir in different situations. Thus, in the swamp the increment is less than in the flat or on the hardwood slope but is more persistent, illustrated by the fact that at the age of 150 years it still continues at an increasing rate. In the swamp the growth of balsam fir resembles more nearly that of spruce. On the hardwood slope the volume growth of balsam fir shows the same tendency as that in height and diameter. It reaches its climax comparatively early and is greatest between the ages of 80 and 95 years. After that it begins to decline. On the flat the maximum rate of volume growth is reached at an age between 50 and 85 years, after which it slightly declines and remains stationary until a very old age.

The increment in cubic feet becomes noticeable much later than the growth in height or diameter; namely, when the tree is 40 or 50 years old. This is still more marked in the case of the increment in board feet. Thus, up to 70 years, even on the hardwood slope, where the growth is fastest, no increment in board feet is noticeable. It reaches its highest rate at the age of 125 years, and continues at a fairly steady rate practically to the limit of the physical life of the tree. On the flat the increment in board feet starts at about 80 and in the swamp at from 90 to 95 years.

TABLE 19.—*Total volume, in cubic feet, of balsam fir in New York, on the basis of diameter and height.*

Diameter breast high (inches).	Height of tree (feet).						
	20	30	40	50	60	70	80
	Total volume (cubic feet).						
3	0.54	0.81					
4	.96	1.43	1.91				
5	1.48	2.21	2.95	3.60			
6		3.15	4.19	5.23	6.24	7.24	
7		4.24	5.63	7.01	8.37	9.72	11.07
8			7.25	9.01	10.76	12.51	14.24
9				11.19	13.38	15.55	17.71
10				13.59	16.23	18.86	21.47
11				16.10	19.25	22.36	25.50
12					22.38	26.06	29.72
13					25.71	29.94	34.14
14					29.12	33.98	38.74
15					32.77	38.14	43.59
16					36.53	42.52	48.59

TABLE 20.—*Total volume, in cubic feet, of balsam fir in Maine, on the basis of diameter and height.*

Diameter breast high (inches).	Height of tree (feet).				
	40	50	60	70	80
	Total volume (cubic feet).				
7	5.68	7.20	8.76	10.34	
8	7.22	9.17	11.12	13.22	15.33
9	8.87	11.26	13.75	16.33	18.98
10		13.20	16.49	19.66	22.91
11		15.77	19.40	23.15	27.03
12			22.38	26.83	31.42
13			25.44	30.58	35.91
14			28.48	34.35	40.45
15			31.52	38.14	45.06
16			34.52	42.03	49.71

TABLE 21.—*Volume growth, in board feet, of balsam fir in New York, on the basis of age.*

Age (years).	Hardwood slope.			Flat.			Swamp.		
	Volume.	Growth in volume every 5 years.	Annual growth in volume within 5-year period.	Volume.	Growth in volume every 5 years.	Annual growth in volume within 5-year period.	Volume.	Growth in volume every 5 years.	Annual growth in volume within 5-year period.
65	22								
70	25	3	0.6						
75	28	3	.6	20					
80	32	4	.8	23	3	0.6			
85	35	3	.6	26	3	.6			
90	39	4	.8	29	3	.6	16		
95	43	4	.8	32	3	.6	18	2	0.4
100	47	4	.8	34	2	.4	20	2	.4
105	51	4	.8	37	3	.6	22	2	.4
110	55	4	.8	40	3	.6	24	2	.4
115	59	4	.8	43	3	.6	26	2	.4
120	63	4	.8	46	3	.6	28	2	.4
125	68	5	1.0	50	4	.8	31	3	.6
130	73	5	1.0	53	3	.6	33	2	.4
135	78	5	1.0	56	3	.6	36	3	.6
140	83	5	1.0	60	4	.8	38	2	.4
145	87	4	.8	63	3	.6	41	3	.6
150	92	5	1.0	66	3	.6	43	2	.4

TABLE 22.—*Volume growth, in cubic feet, of balsam fir in New York, on the basis of age.*

Age (years).	Hardwood slope.			Flat.			Swamp.		
	Volume.	Growth in volume every 5 years.	Annual growth in volume within 5-year period.	Volume.	Growth in volume every 5 years.	Annual growth in volume within 5-year period.	Volume.	Growth in volume every 5 years.	Annual growth in volume within 5-year period.
40	1.73	1.09	0.218	1.09					
45	2.82	1.09	.218	1.92	0.83	0.166	0.98		
50	3.91	1.09	.218	2.80	.88	.176	1.47	0.49	0.098
55	5.00	1.09	.218	3.67	.87	.174	2.01	.54	.108
60	6.10	1.10	.220	4.57	.90	.180	2.60	.59	.118
65	7.20	1.10	.220	5.47	.90	.180	3.21	.61	.122
70	8.30	1.10	.220	6.36	.89	.178	3.88	.67	.134
75	9.40	1.10	.220	7.23	.87	.174	4.56	.68	.136
80	10.50	1.10	.220	8.09	.86	.172	5.24	.68	.136
85	11.63	1.13	.226	8.91	.82	.164	5.92	.68	.136
90	12.74	1.11	.222	9.74	.83	.166	6.59	.67	.134
95	13.85	1.11	.222	10.56	.82	.164	7.26	.67	.134
100	14.93	1.08	.216	11.40	.84	.168	7.94	.68	.136
105	16.00	1.07	.214	12.23	.83	.166	8.62	.68	.136
110	17.05	1.05	.210	13.09	.86	.172	9.30	.68	.136
115	18.10	1.05	.210	13.94	.85	.170	9.98	.68	.136
120	19.15	1.05	.210	14.80	.86	.172	10.66	.68	.136
125	20.19	1.04	.208	15.64	.84	.168	11.34	.68	.136
130	21.23	1.04	.208	16.50	.86	.172	12.02	.68	.136
135	22.29	1.06	.212	17.34	.84	.168	12.70	.68	.136
140	23.34	1.05	.210	18.20	.86	.172	13.38	.68	.136
145	24.37	1.03	.206	19.04	.84	.168	14.09	.71	.142
150	25.43	1.06	.212	19.90	.86	.172	14.80	.71	.142

TABLE 23.—*Volume growth, in cubic feet, of balsam fir in Maine, on the basis of age.*

Age (years).	Volume.	Growth in volume every 5 years.	Annual growth in volume within 5-year period.
65	8.81		
70	10.64	1.83	0.366
75	12.55	1.91	.382
80	14.50	1.95	.390
85	16.51	2.01	.402
90	18.60	2.09	.418
95	20.73	2.13	.426
100	22.90	2.17	.434

TAPER.

Tables 24 to 29 show the taper of balsam fir in different situations in New York and Maine, expressed in inches and in per cent of the diameter breast high. The diameter breast high inside bark is taken as 100, and the diameters inside bark at 10, 20, 30, etc., feet from the ground expressed as fractions.

These taper tables furnish a basis for the construction of volume tables in any log scale or in cubic measure, and serve in general to indicate the development of the bole under various conditions of growth. Thus, they show the more spindling development of balsam fir in the swamp than on either the flat or hardwood slope, and the better development, on the whole, in Maine than in New York.

TABLE 24.—*Taper of balsam fir in New York on swamp.*

[Expressed in per cent of the diameter inside bark breast high.]

Diameter breast high (inches).	Height above ground (feet).						
	4.5	10	20	30	40	50	60
20-foot trees.							
2	100	72.2					
3	100	74.1					
4	100	70.3					
30-foot trees.							
2	100	84.2	52.6				
3	100	85.7	57.1				
4	100	86.5	56.8				
5	100	89.1	58.7				
6	100	87.5	58.9				
7	100	89.2	60.0				
40-foot trees.							
2	100	89.5	68.4	42.1			
3	100	89.3	71.4	42.9			
4	100	91.9	73.0	43.2			
5	100	91.5	72.3	44.7			
6	100	91.1	73.2	46.4			
7	100	92.3	75.4	46.2			
8	100	92.0	74.7	46.7			
9	100	91.8	72.9	47.1			

TABLE 24.—*Taper of balsam fir in New York on swamp*—Continued.

Diameter breast high (inches).	Height above ground (feet).						
	4.5	10	20	30	40	50	60
	50-foot trees.						
4	100	92.1	78.9	63.2	28.9		
5	100	93.6	80.9	61.7	31.9		
6	100	93.0	77.2	59.6	31.6		
7	100	93.9	78.8	59.1	31.8		
8	100	93.3	78.7	58.7	33.3		
9	100	91.8	77.6	57.6	32.9		
10	100	91.5	77.7	57.4	33.0		
11	100	92.2	77.7	57.3	33.0		
12	100	91.2	77.0	56.6	31.9		
13	100	91.0	76.2	56.6	32.0		
	60-foot trees.						
6	.100	94.7	82.5	66.7	47.4	24.6	
7	100	93.9	81.8	66.7	47.0	24.2	
8	100	93.4	81.6	65.8	46.1	23.7	
9	100	92.9	81.2	65.9	45.9	23.5	
10	100	93.6	80.9	66.0	45.7	23.4	
11	100	92.3	80.8	65.4	46.2	23.1	
12	100	92.0	80.5	65.5	46.0	23.9	
13	100	91.1	79.7	65.9	46.3	23.6	
	70-foot trees.						
6	100	94.8	82.8	67.2	50.0	31.0	13.8
7	100	94.0	83.6	68.7	52.2	32.8	16.4
8	100	94.7	82.9	69.7	52.6	35.5	17.1
9	100	94.1	83.5	70.6	55.3	36.5	17.6
10	100	93.7	83.2	70.5	54.7	37.9	18.9
11	100	93.3	82.7	71.2	55.8	38.5	20.2
12	100	93.0	82.5	71.1	56.1	38.6	20.2
13	100	91.9	82.3	71.0	56.5	39.5	21.0

TABLE 25.—*Diameter inside bark, in inches, of balsam fir in New York on swamp, at different heights above the ground.*

[Based on 341 trees.]

Diameter breast high (inches).	Height above ground (feet).						
	4.5	10	20	30	40	50	60
	Diameter inside bark (inches).						
	20-foot trees.						
2	1.8	1.3					
3	2.7	2.0					
4	3.7	2.6					
	30-foot trees.						
2	1.9	1.6	1.0				
3	2.8	2.4	1.6				
4	3.7	3.2	2.1				
5	4.6	4.1	2.7				
6	5.6	4.9	3.3				
7	6.5	5.8	3.9				

TABLE 25.—*Diameter inside bark, in inches, of balsam fir in New York on swamp, at different heights above the ground*—Continued.

Diameter breast high (inches).	Height above ground (feet).						
	4.5	10	20	30	40	50	60
	Diameter inside bark (inches).						
	40-foot trees.						
2	1.9	1.7	1.3	0.8			
3	2.8	2.5	2.0	1.2			
4	3.7	3.4	2.7	1.6			
5	4.7	4.3	3.4	2.1			
6	5.6	5.1	4.1	2.6			
7	6.5	6.0	4.9	3.0			
8	7.5	6.9	5.6	3.5			
9	8.5	7.8	6.2	4.0			
	50-foot trees.						
4	3.8	3.5	3.0	2.4	1.1		
5	4.7	4.4	3.8	2.9	1.5		
6	5.7	5.3	4.4	3.4	1.8		
7	6.6	6.2	5.2	3.9	2.1		
8	7.5	7.0	5.9	4.4	2.5		
9	8.5	7.8	6.6	4.9	2.8		
10	9.4	8.6	7.3	5.4	3.1		
11	10.3	9.5	8.0	5.9	3.4		
12	11.3	10.3	8.7	6.4	3.6		
13	12.2	11.1	9.3	6.9	3.9		
	60-foot trees.						
6	5.7	5.4	4.7	3.8	2.7	1.4	
7	6.6	6.2	5.4	4.4	3.1	1.6	
8	7.6	7.1	6.2	5.0	3.5	1.8	
9	8.5	7.9	6.9	5.6	3.9	2.0	
10	9.4	8.8	7.6	6.2	4.3	2.2	
11	10.4	9.6	8.4	6.8	4.8	2.4	
12	11.3	10.4	9.1	7.4	5.2	2.7	
13	12.3	11.2	9.8	8.1	5.7	2.9	
	70-foot trees.						
6	5.8	5.5	4.8	3.9	2.9	1.8	0.8
7	6.7	6.3	5.6	4.6	3.5	2.2	1.1
8	7.6	7.2	6.3	5.3	4.0	2.7	1.3
9	8.5	8.0	7.1	6.0	4.7	3.1	1.5
10	9.5	8.9	7.9	6.7	5.2	3.6	1.8
11	10.4	9.7	8.6	7.4	5.8	4.0	2.1
12	11.4	10.6	9.4	8.1	6.4	4.4	2.3
13	12.4	11.4	10.2	8.8	7.0	4.9	2.6

TABLE 26.—*Taper of balsam fir in New York on hardwood slope and flat.*

[Expressed in per cent of the diameter inside bark breast high.]

Diameter breast high (inches).	Height above ground (feet).							
	4.5	10	20	30	40	50	60	70
	20-foot trees.							
2	100	77.8						
3	100	78.6						
4	100	78.4						
5	100	78.3						

TABLE 26.—*Taper of balsam fir in New York on hardwood slope and flat—Continued.*

Diameter breast high (inches).	Height above ground (feet).							
	4.5	10	20	30	40	50	60	70
				30-foot trees.				
2	100	84.2	47.4					
3	100	89.3	53.6					
4	100	89.5	60.5					
5	100	91.5	63.8					
6	100	91.2	64.9					
7	100	92.4	68.2					
				40-foot trees.				
2	100	94.7	73.7	47.4				
3	100	92.9	75.0	50.0				
4	100	92.1	73.7	50.0				
5	100	91.7	75.0	50.0				
6	100	93.0	75.4	50.9				
7	100	91.0	74.6	50.7				
8	100	92.1	76.3	52.6				
9	100	91.9	75.6	52.3				
10	100	91.6	76.8	53.7				
				50-foot trees.				
5	100	93.9	81.6	65.3	38.8			
6	100	93.1	81.0	63.8	37.9			
7	100	94.0	80.6	64.2	37.3			
8	100	93.4	81.6	64.5	36.8			
9	100	91.9	80.2	62.8	36.0			
10	100	92.6	80.0	63.2	36.8			
11	100	92.3	79.8	63.5	36.5			
12	100	92.1	78.9	63.2	36.8			
13	100	91.9	79.7	62.6	36.6			
				60-foot trees.				
6	100	94.8	86.2	72.4	53.4	29.3		
7	100	95.5	86.6	71.6	53.7	28.4		
8	100	93.5	83.1	70.1	51.9	27.3		
9	100	94.2	83.7	69.8	51.2	26.7		
10	100	93.7	83.2	69.5	50.5	26.3		
11	100	93.3	81.9	69.5	50.5	25.7		
12	100	93.0	81.6	69.3	50.0	26.3		
13	100	92.7	82.1	69.1	49.6	26.0		
14	100	91.7	81.2	68.4	48.9	25.6		
15	100	91.5	81.7	68.3	48.6	26.1		
16	100	91.4	81.5	68.2	49.0	25.8		
				70-foot trees.				
8	100	94.8	87.0	75.3	61.0	42.9	23.4	
9	100	95.3	86.0	75.6	60.5	43.0	22.1	
10	100	94.7	85.3	74.7	60.0	42.1	22.1	
11	100	93.3	83.8	74.3	59.0	41.0	21.0	
12	100	93.9	84.2	73.7	58.8	40.4	21.1	
13	100	92.7	83.1	73.4	58.9	40.3	21.0	
14	100	92.5	82.7	72.9	58.6	39.8	21.1	
15	100	92.3	82.5	72.7	58.0	39.9	20.3	
16	100	92.1	82.9	73.0	57.9	40.1	20.4	
				80-foot trees.				
9	100	95.3	88.4	80.2	67.4	52.3	34.9	16.3
10	100	94.7	87.4	78.9	66.3	51.6	34.7	16.8
11	100	94.2	86.5	77.9	66.3	51.0	34.6	17.3
12	100	93.9	85.1	76.3	64.9	50.9	34.2	17.5
13	100	93.5	84.7	75.8	64.5	50.0	34.7	17.7
14	100	93.3	83.6	75.4	63.4	50.0	34.3	17.9
15	100	93.0	83.9	74.8	63.6	50.3	35.0	18.2
16	100	92.8	83.0	74.5	63.4	49.7	35.3	18.3

TABLE 27.—*Diameter inside bark, in inches, of balsam fir in New York on hardwood slope and flat, at different heights above the ground.*

[Based on 1,109 trees.]

Diameter breast high (inches).	Height above ground (feet).							
	4.5	10	20	30	40	50	60	70
	Diameter inside bark (inches).							
20-foot trees.								
2	1.8	1.4						
3	2.8	2.2						
4	3.7	2.9						
5	4.6	3.6						
30-foot trees.								
2	1.9	1.6	0.9					
3	2.8	2.5	1.5					
4	3.8	3.4	2.3					
5	4.7	4.3	3.0					
6	5.7	5.2	3.7					
7	6.6	6.1	4.5					
40-foot trees.								
2	1.9	1.8	1.4	0.9				
3	2.8	2.6	2.1	1.4				
4	3.8	3.5	2.8	1.9				
5	4.8	4.4	3.6	2.4				
6	5.7	5.3	4.3	2.9				
7	6.7	6.1	5.0	3.4				
8	7.6	7.0	5.8	4.0				
9	8.6	7.9	6.5	4.5				
10	9.5	8.7	7.3	5.1				
50-foot trees.								
5	4.9	4.6	4.0	3.2	1.9			
6	5.8	5.4	4.7	3.7	2.2			
7	6.7	6.3	5.4	4.3	2.5			
8	7.6	7.1	6.2	4.9	2.8			
9	8.6	7.9	6.9	5.4	3.1			
10	9.5	8.8	7.6	6.0	3.5			
11	10.4	9.6	8.3	6.6	3.8			
12	11.4	10.5	9.0	7.2	4.2			
13	12.3	11.3	9.8	7.7	4.5			
60-foot trees.								
6	5.8	5.5	5.0	4.2	3.1	1.7		
7	6.7	6.4	5.8	4.8	3.6	1.9		
8	7.7	7.2	6.4	5.4	4.0	2.1		
9	8.6	8.1	7.2	6.0	4.4	2.3		
10	9.5	8.9	7.9	6.6	4.8	2.5		
11	10.5	9.8	8.6	7.3	5.3	2.7		
12	11.4	10.6	9.3	7.9	5.7	3.0		
13	12.3	11.4	10.1	8.5	6.1	3.2		
14	13.3	12.2	10.8	9.1	6.5	3.4		
15	14.2	13.0	11.6	9.7	6.9	3.7		
16	15.1	13.8	12.3	10.3	7.4	3.9		

TABLE 27.—*Diameter inside bark, in inches, of balsam fir in New York on hardwood slope and flat, at different heights above the ground*—Continued.

Diameter breast high (inches).	Height above ground (feet).							
	4.5	10	20	30	40	50	60	70
	Diameter inside bark (inches).							
	70-foot trees.							
8	7.7	7.3	6.7	5.8	4.7	3.3	1.8	
9	8.6	8.2	7.4	6.5	5.2	3.7	1.9	
10	9.5	9.0	8.1	7.1	5.7	4.0	2.1	
11	10.5	9.8	8.8	7.8	6.2	4.3	2.2	
12	11.4	10.7	9.6	8.4	6.7	4.6	2.4	
13	12.4	11.5	10.3	9.1	7.3	5.0	2.6	
14	13.3	12.3	11.0	9.7	7.8	5.3	2.8	
15	14.3	13.2	11.8	10.4	8.3	5.7	2.9	
16	15.2	14.0	12.6	11.1	8.8	6.1	3.1	
	80-foot trees.							
9	8.6	8.2	7.6	6.9	5.8	4.5	3.0	1.4
10	9.5	9.0	8.3	7.5	6.3	4.9	3.3	1.6
11	10.4	9.8	9.0	8.1	6.9	5.3	3.6	1.8
12	11.4	10.7	9.7	8.7	7.4	5.8	3.9	2.0
13	12.4	11.6	10.5	9.4	8.0	6.2	4.3	2.2
14	13.4	12.5	11.2	10.1	8.5	6.7	4.6	2.4
15	14.3	13.3	12.0	10.7	9.1	7.2	5.0	2.6
16	15.3	14.2	12.7	11.4	9.7	7.6	5.4	2.8

TABLE 28.—*Taper of balsam fir in Maine.*

[Expressed in per cent of the diameter inside bark breast high.]

Diameter breast high (inches).	Height above ground (feet).								
	4.5	10	20	30	40	50	60	70	80
	40-foot trees.								
6	100	94.7	80.7	50.9					
7	100	91.0	77.6	50.7					
8	100	90.8	76.3	50.0					
9	100	90.6	76.5	49.4					
10	100	89.5	74.7	49.5					
	50-foot trees.								
6	100	94.7	82.5	64.9	36.8				
7	100	92.5	80.6	64.2	35.8				
8	100	92.1	80.3	64.5	36.8				
9	100	91.8	81.2	63.5	37.6				
10	100	91.6	80.0	64.2	37.9				
11	100	91.3	79.8	63.5	39.4				
12	100	92.0	80.5	63.7	39.8				
	60-foot trees.								
6	100	94.8	86.2	75.9	56.9	32.8			
7	100	94.0	85.1	74.6	56.7	32.8			
8	100	94.7	85.5	73.7	55.3	31.6			
9	100	93.0	83.7	72.1	54.7	31.4			
10	100	92.6	83.2	71.6	54.7	30.5			
11	100	93.3	82.7	71.2	53.8	30.8			
12	100	92.1	81.6	69.3	52.6	29.8			
13	100	91.9	80.5	68.3	52.0	29.3			
14	100	91.7	79.5	67.4	52.3	29.5			

TABLE 28.—*Taper of balsam fir in Maine*—Continued.

Diameter breast high (inches).	Height above ground (feet).								
	4.5	10	20	30	40	50	60	70	80
	70-foot trees.								
8	100	96.1	89.5	80.3	68.4	48.7	27.6	....	....
9	100	94.2	87.2	77.9	65.1	47.7	26.7	....	....
10	100	94.7	86.3	76.8	64.2	47.4	26.3	....	....
11	100	93.3	84.8	75.2	62.9	46.7	24.8	....	....
12	100	93.0	84.2	74.6	62.3	45.6	24.6	....	....
13	100	92.7	83.7	74.0	61.8	44.7	23.6	....	....
14	100	92.5	82.7	72.9	60.9	43.6	23.3	....	....
15	100	92.3	81.7	71.8	59.9	43.0	22.5	....	....
16	100	92.1	80.8	70.9	59.6	42.4	21.9	....	....
	80-foot trees.								
8	100	97.4	92.1	84.2	72.4	57.9	40.8	18.4	....
9	100	95.3	89.5	82.6	70.9	57.0	39.5	18.6	....
10	100	94.7	88.4	81.1	69.5	56.8	38.9	18.9	....
11	100	94.3	86.7	79.0	68.6	55.2	38.1	18.1	....
12	100	93.0	86.1	78.3	67.8	54.8	37.4	18.3	....
13	100	93.5	85.5	77.4	67.7	54.8	37.1	18.5	....
14	100	93.2	85.7	77.4	68.4	54.1	36.8	18.8	....
15	100	93.0	84.6	76.9	67.8	53.8	37.1	18.9	....
16	100	92.8	84.2	77.0	67.8	53.9	36.8	19.1	....
	90-foot trees.								
10	100	94.8	88.5	80.2	69.8	58.3	42.7	26.0	11.5
11	100	95.2	88.6	81.0	71.4	59.0	43.8	27.6	12.4
12	100	94.7	88.6	80.7	72.8	60.5	45.6	28.9	13.2
13	100	94.4	87.9	80.6	72.6	61.3	46.0	29.8	13.7
14	100	94.0	87.3	80.6	73.1	61.9	47.0	29.9	14.9
15	100	93.8	86.8	80.6	73.6	62.5	47.9	30.6	14.6
16	100	93.4	86.9	81.0	74.5	63.4	49.0	32.0	15.7

TABLE 29.—*Diameter inside bark, in inches, of balsam fir in Maine, at different heights above the ground.*

[Based on 885 trees.]

Diameter breast high (inches).	Height above ground (feet).								
	4.5	10	20	30	40	50	60	70	80
	Diameter inside bark (inches).								
	40-foot trees.								
6	5.7	5.4	4.6	2.9	....	....	....	....	....
7	6.7	6.1	5.2	3.4	....	....	....	....	....
8	7.6	6.9	5.8	3.8	....	....	....	....	....
9	8.5	7.7	6.5	4.2	....	....	....	....	....
10	9.5	8.5	7.1	4.7	....	....	....	....	....
	50-foot trees.								

TABLE 29.—*Diameter inside bark, in inches, of balsam fir, in Maine, at different heights above the ground*—Continued.

Diameter breast high (inches).	Height above ground (feet).								
	4.5	10	20	30	40	50	60	70	80
	Diameter inside bark (inches).								
	60-foot trees.								
6	5.8	5.5	5.0	4.4	3.3	1.9			
7	6.7	6.3	5.7	5.0	3.8	2.2			
8	7.6	7.2	6.5	5.6	4.2	2.4			
9	8.6	8.0	7.2	6.2	4.7	2.7			
10	9.5	8.8	7.9	6.8	5.2	2.9			
11	10.4	9.7	8.6	7.4	5.6	3.2			
12	11.4	10.5	9.3	7.9	6.0	3.4			
13	12.3	11.3	9.9	8.4	6.4	3.6			
14	13.2	12.1	10.5	8.9	6.9	3.9			
	70-foot trees.								
8	7.6	7.3	6.8	6.1	5.2	3.7	2.1		
9	8.6	8.1	7.5	6.7	5.6	4.1	2.3		
10	9.5	9.0	8.2	7.3	6.1	4.5	2.5		
11	10.5	9.8	8.9	7.9	6.6	4.9	2.6		
12	11.4	10.6	9.6	8.5	7.1	5.2	2.8		
13	12.3	11.4	10.3	9.1	7.6	5.5	2.9		
14	13.3	12.3	11.0	9.7	8.1	5.8	3.1		
15	14.2	13.1	11.6	10.2	8.5	6.1	3.2		
16	15.1	13.9	12.2	10.7	9.0	6.4	3.3		
	80-foot trees.								
8	7.6	7.4	7.0	6.4	5.5	4.4	3.1	1.4	
9	8.6	8.2	7.7	7.1	6.1	4.9	3.4	1.6	
10	9.5	9.0	8.4	7.7	6.6	5.4	3.7	1.8	
11	10.5	9.9	9.1	8.3	7.2	5.8	4.0	1.9	
12	11.5	10.7	9.9	9.0	7.8	6.3	4.3	2.1	
13	12.4	11.6	10.6	9.6	8.4	6.8	4.6	2.3	
14	13.3	12.4	11.4	10.3	9.1	7.2	4.9	2.5	
15	14.3	13.3	12.1	11.0	9.7	7.7	5.3	2.7	
16	15.2	14.1	12.8	11.7	10.3	8.2	5.6	2.9	
	90-foot trees.								
10	9.6	9.1	8.5	7.7	6.7	5.6	4.1	2.5	1.1
11	10.5	10.0	9.3	8.5	7.5	6.2	4.6	2.9	1.3
12	11.4	10.8	10.1	9.2	8.3	6.9	5.2	3.3	1.5
13	12.4	11.7	10.9	10.0	9.0	7.6	5.7	3.7	1.7
14	13.4	12.6	11.7	10.8	9.8	8.3	6.3	4.0	2.0
15	14.4	13.5	12.5	11.6	10.6	9.0	6.9	4.4	2.1
16	15.3	14.3	13.3	12.4	11.4	9.7	7.5	4.9	2.4

CUBIC FOOT VOLUME.

Tables 30 to 32 give the merchantable volume of trees of different diameters and heights inside the bark in cubic feet, cutting to a top-diameter limit of 4 inches in Maine and New York and to 6 inches in New Hampshire.

TABLE 30.—*Merchantable volume of balsam fir in New York, in cubic feet inside the bark, on basis of diameter and height.*

[Average top diameter, 4 inches; based on 947 trees.]

Diameter breast high (inches).	Height of tree (feet).				
	40	50	60	70	80
	Merchantable volume (cubic feet).				
6	3.5	4.0			
7	4.5	5.1	5.9		
8	5.9	6.8	8.0	9.1	
9	7.6	8.9	10.4	11.9	
10		11.2	13.0	14.8	16.6
11		13.6	15.6	17.6	19.8
12			18.3	20.9	23.6
13			21.1	24.5	27.7
14				28.4	32.2
15				32.9	37.5
16				37.8	43.2

TABLE 31.—*Merchantable volume of balsam fir in Maine, in cubic feet inside the bark, on basis of diameter and height.*

[Average top diameter, 4 inches; based on 330 trees.]

Diameter breast high (inches).	Height of tree (feet).				
	50	60	70	80	90
	Merchantable volume (cubic feet).				
8	7.7	9.3	10.7	12.2	
9	9.4	11.3	13.1	14.9	
10	11.3	13.7	15.9	18.1	
11	13.5	16.4	19.1	21.6	23.6
12		19.4	22.4	25.4	27.7
13		22.9	26.1	29.4	32.4
14				34.0	37.8
15				39.0	44.1

TABLE 32.—*Merchantable volume of balsam fir in Grafton County, N. H., in cubic feet inside the bark, on basis of diameter and height.*

[Top diameter, 6 inches.]

Diameter breast high (inches).	Height of tree (feet).		
	40	50	60
	Merchantable volume (cubic feet).		
6	1.9	2.1	
7	3.9	4.4	5.0
8	6.0	6.8	7.8
9	8.3	9.5	10.9
10	10.8	12.3	14.2
11	13.5	15.3	17.6
12		18.5	21.2
13		23.7	25.0
14			28.8
15			32.9

CORD VOLUMES.

Tables 33 to 35 give the merchantable volume of trees of different diameters and heights, in cords, for New York, Maine, and New Hampshire. Table 36 gives the number of trees of different heights and diameters per cord for Maine and New York.

In New Hampshire the top diameter is 6 inches and in New York and Maine 4 inches.

TABLE 33.—*Total volume of balsam fir in Maine and New York, in cords, of trees of different diameters and heights.*

[Based on 2,171 trees.]

Diameter breast high (inches).	Height of tree (feet).						
	20	30	40	50	60	70	80
	Cords per tree.						
3	0.005	0.008					
4	.009	.016	0.022				
5	.016	.024	.033	0.042			
6		.034	.045	.057	0.068		
7		.045	.060	.075	.089	0.105	
8			.078	.096	.114	.137	0.168
9			.099	.119	.142	.171	.203
10			.121	.146	.173	.204	.240
11			.146	.178	.206	.241	.278
12				.213	.244	.281	.320
13					.284	.324	.368
14						.366	.419

TABLE 34.—*Merchantable volume of balsam fir in Grafton County, N. H., in cords, of trees of different diameters and heights.*

[Top diameter, 6 inches.]

Diameter breast high (inches).	Height of tree (feet).		
	40	50	60
	Cords per tree.		
6	0.019	0.022	
7	.040	.045	0.052
8	.062	.071	.081
9	.086	.099	.113
10	.112	.128	.148
11	.140	.159	.184
12		.192	.221
13		.247	.260
14			.300
15			.342

TABLE 35.—*Number of trees per cord of balsam fir in Maine and New York.*

[Based on 2,171 trees.]

Diameter breast high (inches).	Height of tree (feet).						
	20	30	40	50	60	70	80
	Trees per cord.						
3	200.0	125.0					
4	111.1	62.5	45.5				
5	62.5	41.7	30.3	23.8			
6		29.4	22.2	17.5	14.7		
7		22.2	16.7	13.3	11.2	9.5	
8			12.8	10.4	8.8	7.3	6.0
9			10.1	8.4	7.0	5.8	4.9
10			8.3	6.8	5.8	4.9	4.2
11			6.8	5.6	4.9	4.1	3.6
12				4.7	4.1	3.6	3.1
13					3.5	3.1	2.7
14						2.7	2.4

Table 36, based on actual measurements of over 7^3 cords, gives the relation between number of trees, number of sticks, and the solid contents of a stacked cord.

TABLE 36.—*Relation between number of trees, number of sticks, and solid contents of a stacked cord.*

[Based on 56.6 cords.]

Average number of trees per cord.	Average number of 4-foot sticks per cord.	Average number of cubic feet per cord.	Average number of trees per cord.	Average number of 4-foot sticks per cord.	Average number of cubic feet per cord.
9.0	85.3	89.72	7.0	65.7	94.56
8.6	74.8	95.79	7.9	72.8	92.22
8.2	74.9	96.06	7.3	71.6	94.64
7.5	63.6	85.62	9.6	87.5	95.84
9.9	83.7	95.84			

If the middle diameter of the average 4-foot sticks in a stacked cord of wood is known, the solid contents of the cord can be readily ascertained, as it varies with the middle diameter of the average stick in the following manner:

TABLE 37.—*Relation between solid contents of a stacked cord and middle diameter of average stick.*

Middle diameter of average stick (inches).	Cubic feet per cord.
6.5	89
7.0	92
7.5	94
8.0	96
8.5	98

CONVERTING FACTORS OF STACKED MEASURE INTO CUBIC FEET.

The following factors can be used for converting stacked measure into cubic measure:

For cords made up of billets 4 feet long and from 4 to 7 inches in diameter, 0.72.

For cords made up of billets 4 feet long and from 8 to 12 inches in diameter, 0.76.

In order to convert a cord of ordinary pulpwood into cubic measure, 128 feet should be multiplied by 0.74, the average converting factor for pulpwood.

FORM FACTORS.

Tables 38 and 39 give breast-high form factors for Maine and for New York.

TABLE 38.—*Form factors (breast high) for New York.*

Diameter breast high (inches).	Height of tree (feet).						
	20	30	40	50	60	70	80
	Form factor.						
3	0.552	0.551					
4	.547	.546	0.546				
5	.542	.541	.541	0.539			
6		.535	.534	.533	0.530	0.527	
7		.529	.527	.525	.522	.520	0.518
8			.519	.518	.514	.512	.510
9				.507	.505	.503	.501
10				.493	.496	.494	.492
11				.488	.486	.484	.483
12					.475	.474	.473
13					.465	.464	.463
14					.454	.454	.453
15					.445	.444	.444
16					.436	.435	.435

TABLE 39.—*Form factors (breast high) for Maine.*

Diameter breast high (inches).	Height of tree (feet).				
	40	50	60	70	80
	Form factor.				
7	0.531	0.539	0.546	0.553	
8	.517	.525	.531	.541	0.549
9	.502	.510	.519	.528	.537
10		.484	.504	.515	.525
11		.473	.490	.501	.512
12			.475	.488	.500
13			.460	.474	.487
14			.444	.459	.473
15			.428	.444	.459
16			.412	.430	.445

BARK VOLUMES.

Table 40 gives the volume of bark of trees of different diameters and heights. On the whole, bark makes up about 10.6 per cent of the total volume of the tree.

TABLE 40.—*Volume of bark for trees of different diameters and heights in Maine.*

Diameter breast high (inches).	Height of tree (feet).				
	40	50	60	70	80
	Volume of bark (cubic feet).				
7	0.6	0.7	0.9	1.0	
8	.7	.9	1.1	1.3	1.5
9	.9	1.1	1.4	1.6	1.9
10		1.3	1.6	2.0	2.3
11		1.6	1.9	2.3	2.7
12			2.2	2.7	3.1
13			2.5	3.1	3.6
14			2.8	3.4	4.0
15			3.2	3.8	4.5
16			3.5	4.2	5.0

BOARD-FOOT VOLUMES.

Tables 41 and 42 give the merchantable volume in board feet (Scribner, Dimick, Maine, and Bangor rules) for trees of different diameters and heights.

TABLE 41.—*Volume of trees of different diameters and heights—Scribner Decimal C and Dimick rules.*

[Based on taper curves, scaled as 8 and 16 foot logs. Stump height assumed, 1 foot.]

SCRIBNER DECIMAL C.

Diameter breast high (inches).	Swamp.				Diameter inside bark of top (inches).	Hardwood slope and flat.					Diameter inside bark of top (inches).
	Height of tree (feet).					Height of tree (feet).					
	40	50	60	70		40	50	60	70	80	
	Volume (board feet).					Volume (board feet).					
7	14	17	20	22	5.8	13	19	27			5.8
8	19	23	27	32	5.9	21	26	33	40		5.9
9	24	31	37	44	6.1	29	34	41	48	56	6.0
10		39	48	57	6.2	38	45	52	60	70	6.1
11		48	60	73	6.4		56	65	75	86	6.2
12		57	73	92	6.6		69	80	92	107	6.3
13		66	87	110	6.8		82	95	111	130	6.4
14								111	132	155	6.4
15								127	153	182	6.5
16								144	174	209	6.6

DIMICK.

Diameter breast high (inches).	Volume standards.					Volume standards.					
7	0.09	0.10	0.12	0.15	5.8	0.11	0.13	0.15			5.8
8	.13	.15	.18	.21	5.9	.15	.18	.21	0.24		5.9
9	.17	.21	.24	.28	6.1	.19	.23	.28	.33	0.39	6.0
10		.26	.31	.36	6.2	.24	.29	.35	.41	.48	6.1
11		.32	.38	.47	6.4		.36	.43	.50	.58	6.2
12		.38	.46	.59	6.6		.44	.51	.59	.68	6.3
13		.44	.54	.71	6.8		.52	.60	.70	.81	6.4
14								.71	.82	.94	6.4
15								.82	.94	1.09	6.5
16								.94	1.07	1.25	6.6

TABLE 42.—*Volume of trees of different diameters and heights—Bangor and Maine rules.*

BANGOR RULE.[1]

Diameter breast high (inches).	Height of tree (feet).						Diameter inside bark of top (inches).
	40	50	60	70	80	90	
	Volume (board feet).						
7	15	21	27	....	....	....	5.9
8	23	29	38	50	64	....	6.2
9	32	40	49	62	78	....	6.4
10	42	51	62	76	93	109	6.6
11	....	63	77	94	110	129	6.8
12	....	77	94	113	132	154	7.0
13	....	....	111	135	160	185	7.1
14	....	....	129	159	191	223	7.2
15	....	....	....	186	226	268	7.3
16	....	....	....	215	262	317	7.4

MAINE RULE.[2]

Diameter breast high (inches).	Height of tree (feet).						Diameter inside bark of top (inches).
7	15	21	26	....	....	....	5.9
8	24	32	40	50	60	....	6.0
9	34	44	54	65	76	....	6.1
10	46	57	69	81	94	106	6.2
11	....	71	85	99	114	129	6.2
12	....	86	102	120	138	157	6.2
13	....	....	121	143	167	193	6.3
14	....	....	140	166	200	236	6.3
15	....	....	....	191	236	283	6.3
16	....	....	....	215	271	333	6.4

[1] Based on taper curves, scaled as 16-foot logs. Stump height assumed, 1 foot.
[2] Based on taper curves, scaled as 8 and 16 foot logs. Stump height assumed, 1 foot.

RATIO BETWEEN BOARD AND CUBIC MEASURE.

Table 43 gives the ratio between board measure (Maine rule) and cubic feet.

TABLE 43.—*Ratio between board measure (Maine rule) and cubic feet (merchantable contents).*

[Based on 330 trees.]

Diameter breast high (inches).	Board feet per cubic foot.	Diameter breast high (inches).	Board feet per cubic foot.
7	3.0	12	3.4
8	3.2	13	3.4
9	3.3	14	3.5
10	3.3	15	3.5
11	3.4	16	3.5

CLEAR LENGTH AND USED LENGTH.

Tables 44 and 45 give the clear length and the used length of trees of different diameters in Maine and New York.

TABLE 44.—*Clear length and used length of balsam fir of different heights and diameters in New York.*

Diameter breast high (inches).	Hardwood slope.			Flat.			Swamp.		
	Total height.	Clear length.[1]	Used length.[2]	Total height.	Clear length.[3]	Used length.[4]	Total height.	Clear length.[5]	Used length.[6]
	Feet.	*Feet.*	*Feet.*	*Feet.*	*Feet.*	*Feet.*	*Feet.*	*Feet.*	*Feet.*
6	48	19	27	48	24	26	45	17	25
7	53	22	31	52	25	30	49	21	28
8	57	24	35	56	26	33	52	23	31
9	61	26	38	60	27	37	55	24	33
10	63	27	42	62	28	40	58	25	36
11	66	28	44	65	29	43	60	25	39
12	69	29	47	67	29	46	62	25	43
13	71	30	49	70	30	48	64	25	46
14	73	31	51	72	31	50	66		
15	75	32	52	74	33	52			
16	78	32	54	76	34	53			

[1] Based on 440 trees.   [3] Based on 386 trees.   [5] Based on 344 trees.
[2] Based on 560 trees.   [4] Based on 333 trees.   [6] Based on 202 trees.

TABLE 45.—*Clear length and used length of balsam fir of different heights and diameters in Maine.*

[Clear length based on 407 trees; used length based on 379 trees.]

Diameter breast high (inches).	Total height.	Clear length.	Used length.	Diameter breast high (inches).	Total height.	Clear length.	Used length.
	Feet.	*Feet.*	*Feet.*		*Feet.*	*Feet.*	*Feet.*
6	52			12	73	40	29
7	57	38	22	13	75	40	30
8	62	39	23	14	77	40	31
9	65	39	25	15	79	40	32
10	68	40	26	16	80	40	33
11	71	40	27	17	82	40	34

PER CENT OF CULL AND WASTE.

The average cull within merchantable dimensions, that is, for the portion of the trees from stump to 4-inch top, constitutes on the average about 11.2 per cent of the merchantable yield. The top and stump form about 8.4 per cent of the total volume; the bark, 10.6 per cent. In other words, about 19 per cent of the total volume of the tree at present remains unutilized. Of the remaining merchantable part of the tree, 11.2 per cent must be allowed for cull.

The yield of balsam fir fluctuates within wide limits. Since it grows with spruce and other species, its yield naturally depends upon the degree of admixture. An idea of what can be expected from balsam fir may best be formed from pure stands in the swamps or flats. For New York a good average for large flats, cutting for pulp to 7 inches diameter breast high, is 15 cords to the acre. Exceptional areas have cut as high as 40 cords. In swamps, while the stands are usually dense, the individual trees are of small size, and the yield per acre on the whole is smaller than on the flats. Ten cords to the acre may be considered a good average. On the hardwood slope the yield varies more than for any other type; on an average it runs about 7 cords to the acre.

In Maine the yield runs much higher than in New York. Pure stands of balsam fir on flats will yield, as a general rule, about 25 cords to the acre and occasionally as high as 30 cords for stands from 70 to 100 years old. On the hardwood slope the yield is only half of that on the flat, about 12.5 cords to the acre.

Tables 46 and 47 give the results of actual measurements of yield in the Adirondacks and in Maine.

TABLE 46.— *Yield of balsam fir in New York, based on 10 sample plots, covering an area of 9 acres.*

SWAMP.

Average age of merchantable stand (years).	Total yield per acre.		Average number of merchantable trees per acre.	Mean annual increment per acre.	
	Cubic feet.	*Cords.*		*Cubic feet.*	*Cords.*
80	922	10.2	88	11.5	½

FLAT.

90	1,270	13.2	102	14.1	
90	1,312	15.3	110	14.6	½
90	1,443	15.0	140	16.0	½
Average	1,342	14.4	117	14.9	½

HARDWOOD SLOPE.

70	444	4.6	36	6.3	
70	685	7.1	60	9.8	
70	760	8.0	49	10.9	
80	713	7.7	55	8.9	
70	607	6.4	46	8.7	
70	928	9.7	86	13.3	
Average	690	7.3	55	9.6	1/10

TABLE 47.— *Yield of balsam fir in Maine, based on 22 sample plots, covering an area of 6 acres.*

FLAT.

Plot No.	Age of stand.	Trees cut per acre.	Total yield.	Yield without bark.	Merchantable volume.		Mean annual increment per acre.
			Cubic feet.	*Cubic feet.*	*Cubic feet.*	*Cords.*	*Cubic feet.*
1................	90	152	2,395	2,156	2,149	22.6	26.6
2................	90	144	2,763	2,501	2,460	25.9	30.7
3................	70	160	2,543	2,268	2,330	24.5	36.3
4................	70	100	1,357	1,212	1,239	13.0	19.4
5................	80	168	2,937	2,614	2,704	28.4	36.7
6................	90	200	3,739	3,364	3,504	36.8	41.5
7................	90	160	3,010	2,674	3,010	31.6	33.4
8................	90	132	2,936	2,627	2,675	28.1	32.6
9................	90	136	2,789	2,473	2,512	26.4	31.0
10................	90	86	1,408	1,261	1,259	13.2	15.6
11................	70	144	2,811	2,435	2,590	28.3	40.1
12................	90	120	2,115	1,896	1,910	20.0	23.5
13................	90	168	2,689	2,388	2,497	26.2	29.8
Average....		144	2,575			25.0	[1] 30.5

HARDWOOD SLOPE.

14................	90	80	1,666	1,490	1,547	16.2	18.5
15................	80	88	1,342	1,214	1,240	13.0	16.8
16................	80	60	1,034	930	957	10.0	12.9
17................	80	68	963	868	881	9.2	12.0
18................	90	96	1,861	1,668	1,683	17.7	20.6
19................	90	122	2,165	1,945	1,933	20.3	24.0
20................	80	48	534	485	480	5.0	6.6
21................	100	62	1,170	1,040	1,050	11.0	11.7
22................	90	52	1,226	1,085	1,112	11.7	13.6
Average....		75	1,329			12.6	[2] 15.2

[1] Equals one-third cord.   [2] Equals one-sixth cord.

OVER LARGE AREAS.

The figures in Tables 46 and 47 represent the yield of carefully selected small areas of balsam-fir stands. Over large areas, including all types of land, the yield is much smaller. The results of measurements of nearly 60,000 acres in three townships of Hamilton County, N. Y., gave an average yield per acre for all types of coniferous lands of 4.4 standards, or 1.5 cords (Table 48).

48.—*Average yield of balsam fir over large areas in Hamilton County, N. Y.*

[Cutting to a limit of 10 inches and over in diameter breast high.]

Type.	Area (acres).	Total yield (standards).	Average yield per acre (standards).
ip 5:			
mp and spruce land	10,376	39,676.58	3.82
Do	4,405	22,677.90	5.15
Do	2,072	10,675.35	5.15
Total	16,853	73,029.83	
Average per acre			4.30
ip 41:			
mp and spruce land	6,960	19,585.20	2.81
Do	1,869	6,397.76	3.42
Do	10,982	57,755.88	5.26
Total	19,811	83,738.84	
Average per acre			4.20
ip 6:			
mp and spruce land	122	52.92	.43
Do	12,156	55,374.84	4.56
Do	3,609	15,618.26	4.33
Do	2,779	18,391.98	6.62
Do	1,332	5,740.92	4.31
Do	1,464	4,832.88	3.30
Total	21,462	100,011.80	
Average per acre			4.70

asurements of nearly 17,000 acres in Herkimer County, N. Y., an average yield for all types of coniferous land (both virgin and ver) of only 1.4 standards. The yield for swamp land, which is y balsam-fir land, ran on an average as high as 5.42 standards, irly 2 cords to the acre. (Table 49.)

LE 49.—*Average yield of balsam fir over large areas in Herkimer County, N. Y.*

[Cutting to a limit of 10 inches in diameter breast high.]

Type.	Area (acres).	Total yield (standards).	Average yield per acre (standards).
and:			
in	3,732	4,105	1.10
over	4,158	5,073	1.22
ital	7,890	9,178	1.16
in	210	1,187	5.65
over	161	824	5.12
ital	371	2,011	5.42
and total	8,261	11,189	
verage			1.40

In New Hampshire, measurements of over 2,000 acres gave an average yield of 482 board feet for balsam and 1,772 board feet for spruce (New Hampshire rule), or nearly 0.8 of a cord of balsam fir per acre, forming about 27 per cent of the entire spruce yield. (Table 50.)

TABLE 50.—*Average yield of balsam fir over large areas in Grafton County, N. H.*

[New Hampshire log rule.]

Area (acres).	Total yield.		Average yield per acre.	
	Spruce.	Balsam.	Spruce.	Balsam.
	Board feet.	*Board feet.*	*Board feet.*	*Board feet.*
107	124,976	63,879	1,168	597
112	273,840	93,520	2,445	835
115	94,645	10,925	823	95
135	170,235	14,715	1,261	109
141	38,211	846	271	6
190	651,510	67,260	3,429	354
574	1,402,856	392,616	2,444	684
259	437,192	172,494	1,688	666
446	617,264	174,386	1,384	391
154	146,146	84,700	949	550
Total (2,233)	3,956,875	1,075,341	1,772	482

INCREMENT.

The sample plots in New York and Maine (Tables 46 and 47) showed that mature stands of balsam fir produce annually from one-sixth to one-third of a cord of wood per acre. At such a rate the poorest land produces 10 cords per acre in 60 years, and the better land 10 cords of pulpwood every 30 years. This annual increment is very low as compared with the yields obtainable under forest management. The increment should be at least two-thirds of a cord, or possibly 1 cord a year.

MANAGEMENT.

EFFECT OF PAST CUTTING.

Balsam fir is so closely associated with spruce wherever it occurs that it is impossible to outline a system of management for one species that will not at the same time affect the other. Both species are almost constantly contesting for the occupancy of the ground. If left to themselves the greater tolerance and more persistent growth of spruce would undoubtedly in the long run secure for it the predominance in the present forests as they formerly did in the virgin stands, before the interference of man. Lumbering, however, has turned the scale of the struggle between the different species in favor of trees of smaller commercial importance. Thus, white pine, the most valuable species of the northeastern forests, was taken first, with the result that it was unable to hold its own against its competitors. Then came the turn of spruce. The latter, in many cases, is now being cut for the third time, smaller logs being taken at each new lumbering. Balsam, on the other hand, has been spared until recently and thus given a chance to spread at the expense of spruce. These facts are well brought out by measurements taken in Maine on 20 acres of virgin and 20 acres of forest cut over once. The difference in the representation of the two species in virgin and cut-over forest is especially striking in the trees of small diameters, since not enough time has elapsed after cutting to affect in any great degree the large trees.

TABLE 51.—*Average number of spruce and balsam fir up to 12 inches in diameter breast high on an acre of virgin and cut-over forest in Maine.*

Diameter breast high (inches).	Virgin.		Cut-over.	
	Spruce.	Balsam.	Spruce.	Balsam.
2	29.2	12.8	13.0	35.2
3	51.0	12.0	8.6	23.4
4	44.2	11.8	7.4	21.4
5	39.4	8.6	5.4	19.8
6	24.4	7.0	4.4	17.6
7	23.2	6.4	4.2	14.6
8	17..8	6.4	4.0	13.0
9	17.4	2.2	3.4	10.2
10	14.2	2.0	3.0	9.8
11	9.2	1.6	2.6	4.6
12	6.4	1.2	2.6	4.0
Total	276.4	72.0	58.6	173.6

The rapid spreading of balsam over cut or burnt spruce land is due chiefly to its prolific seeding, love for light, and rapid growth. In this respect, as in many others, balsam occupies the same place among the northeastern conifers as aspen does among the deciduous species. It is the first of all conifers to take possession of openings, burnt or cut-over land, and at present outnumbers spruce in the young growth and smaller diameters throughout the northern woods. Table 52 shows the number of spruce and fir trees on an average acre, based on actual measurements of 955 acres in the forests of Maine. The measurements were taken on the slope type, where spruce is more at home than is balsam.

TABLE 52.—*Number of spruce and fir trees on an average acre, based on 952 acres in Maine.*

Diameter breast high (inches).	Spruce.	Balsam fir.	Diameter breast high (inches).	Spruce.	Balsam fir.
2	9.1	17.8	10	4.0	6.1
3	9.9	18.7	11	3.6	3.9
4	9.1	18.0	12	3.4	2.3
5	7.2	15.9			
6	5.8	13.2	Total	66.1	124.6
7	5.4	11.2			
8	4.6	9.6	Trees, 2 inches to 8 inches, inclusive	51.1	104.4
9	4.0	7.9			

SILVICULTURAL SYSTEMS OF CUTTING.

Upon the methods of cutting adopted in spruce stands will depend whether the future forest will be chiefly spruce or balsam or whether there will be future growth at all. In discussing these methods the economic limitations and specific conditions which may affect their application are not considered. These must necessarily differ for each particular forest tract. In a general discussion of the silvicultural system adapted to spruce and balsam it is possible to lay down only general principles.

Natural reproduction may be secured in spruce-balsam fir stands by two methods: (1) Clear cutting and (2) gradual cuttings.

CLEAR CUTTING WITH NATURAL REPRODUCTION.

In clear cutting, natural reproduction from stands adjoining the cutting must be relied upon to restock the area. The size and form of the clear-cut areas are therefore factors in the success of the reproduction. If natural reproduction is desired, the greatest width of the

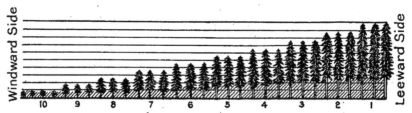

FIG. 6.—Results secured by logging on the leeward side of balsam fir-spruce stands. The youngest stands are found on the windward side and deflect the wind upward, preventing windfall among the older trees.

area to be cut clear in spruce-balsam fir stands should not exceed double the height of the adjoining stand from which reseeding is expected. For example, if the average height of a spruce and balsam-fir stand is 75 feet, then the width of the area which is to be cut clear should not be greater than 150 feet. The length of the area does not affect the natural reproduction and should depend, therefore, upon the amount of timber to be cut, convenience of logging, and similar considerations. In general, then, clear cutting with natural reproduction in spruce-balsam fir stands should take the form of long narrow strips.

Since both spruce and balsam are shallow-rooted trees and therefore subject to windfall, logging operations should as far as possible always begin on the leeward side of the mature timber, and proceed against the wind. If logging were to begin on the windward side there would always be danger from windfall in the stands adjoining the logged area. When the entire forest is cut over in this way, the youngest stands will be on the windward side, their tops forming a gradual ascending plane (fig. 6). The wind is thus deflected

pward, without breaking into the older stands. Logging from the
eward side also permits the seed to be carried by the wind from the
nature stands to the logged-over area.

Successive strips.—No matter how narrow the strips are made,
hey should not be cut one after another every year, unless there is
afficient young growth to insure a full stand. Spruce and balsam
o not bear seed every year, but at intervals of from four to six years.
f the strips are cut one after another every year, the logged areas
ould not be reproduced for lack of seed. The stand adjoining
he logged area should be cut only after the latter has been fully
seeded, or at the end of four to six years. With this method of
utting the logging will have to be scattered over a fairly wide
erritory.

Alternate strips.—To avoid too great a scattering of the cuttings,
which necessarily increase the cost of logging, the strips may be cut

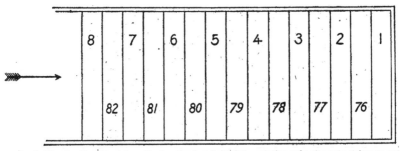

IG. 7.—Cutting in alternate strips. During the first half of the rotation only alternate strips are cut.
The remaining strips are cut over during the second half of the rotation. At the time the remaining
strips are cut the first strips are 75 years old and are capable of reseeding the adjoining clearings.

lternately instead of one after another, at an interval of from
our to six years. In applying this method, the entire tract is divided
nto strips narrow enough to insure natural reproduction. The
ract is cut over twice. The first time only alternate strips are cut;
he second time, the remaining strips. Every year as many strips
nay be cut as are needed to secure the desired amount of timber.
Under this method the timber tract, after it has been entirely cut
ver, would consist of strips of timber in which two adjoining strips
would differ in age by as many years as it took to cut over all of the
lternate strips. If 150 years is decided upon as the rotation for a
mixed stand of spruce and balsam fir, the entire tract would be cut
ver in 150 years, and the alternate strips would be cut over within
he first 75 years. The strips that were cut first would then be 75
ears old when the adjoining strips are cut. At the age of 75 years
oth spruce and balsam bear seed prolifically, and will readily reseed
he adjacent clearings made by cutting the remaining strips (fig. 7).

Cutting in alternate strips tends to concentrate logging, since as
nuch timber may be cut per acre as under the present methods of

culling the forest of trees of certain diameters. It requires, however, great regularity and exactness in logging operations, and may therefore present difficulties, although it is being practiced to a considerable extent in private and State spruce forests abroad.

Thinned and partially cleared strips.—Another modification of the system of clear cutting in strips is especially applicable to large stands of mature timber. Watersheds, or other logging units large enough to allow logging on the same area for a period of five or six years, are divided into strips, all of which are cut over within the five or six year period, but only for two-thirds of their full width. Thus, if the width of the strips is 150 feet, each strip is cut only 100 feet. On the remaining 50 feet of each strip the timber is merely thinned (fig. 8). As logging operations on the area will go on for five or six years, there should be one or two good seed years during which the logged areas will be reproduced from the adjoining 50-foot

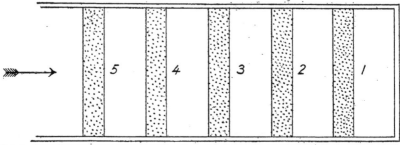

Fig. 8.—Partially cleared and thinned strips. Each strip is cut only for two-thirds of its width. On the remaining one-third the timber is only thinned. Reproduction takes place on the adjoining clearings and under the thinned stand. When reproduction is secured, the remaining one-third of the strip is cut clean. The entire logging area is reproduced within five or six years.

strips of timber. Since 50-foot strips are thinned, reproduction will occur on them. As soon as young growth appears on the clearings and under the trees left uncut, the 50-foot strips are also taken, and the entire area is thus cut and reproduced within a few years. This method of cutting is simple and, under favorable conditions, practical as a logging proposition. The great danger is from windfall, to which thinned stands are particularly susceptible.

While often, as in cutting for pulp, clear cutting in strips is the best method, even with the greatest precautions cleared strips often fail to reproduce naturally with the desired species. No matter what modification of the system is practiced, the narrower the strip the greater are the chances for successful natural reproduction. When abundant young growth exists under the old trees, clear cutting need not be in the form of strips, but may cover the entire area bearing reproduction.

Clear cutting in strips must naturally lead to an increase of balsam in the second growth, since it is a prolific seeder and requires more light than does spruce. This is especially true in the case of alternate or successive strips. With partially cleared and thinned strips, however, which are cut practically at the same time, the reproduction of balsam fir can be reduced in favor of spruce if thinning is confined largely or exclusively to balsam fir, thus decreasing its participation in reseeding the ground.

CLEAR CUTTING, WITH ARTIFICIAL REPRODUCTION.

Still another silvicultural method to which both spruce and balsam fir are adapted, particularly for pulp, is clear cutting, with subsequent planting. Such a system, however, presupposes intensive management and a considerable initial outlay of money. The planting of red spruce and balsam fir would be hardly advisable for both silvicultural and financial reasons, because of the former's extremely slow growth and the latter's comparatively inferior qualities. If planting is to be done, it would be better to use more valuable and promising species, such as Norway or possibly white spruce. The cost of establishing a stand artificially is the same whether valuable or inferior species are used. For these reasons clear cutting, with artificial reproduction, would hardly be a profitable undertaking, at least for the balsam fir. The justification for retaining balsam fir in the future stands must be in the ease with which it can be reproduced naturally and cheaply.

GRADUAL CUTTING.

Selection in groups.—Spruce stands are best managed by gradual cuttings. This is essentially the method used in the old-time logging operations, when only the largest trees could be used, and is in vogue now on a number of large spruce tracts owned by pulp and paper companies. Only the larger mature trees or trees of a certain character are taken, and the rest left on the ground for future logging.

Natural reproduction of spruce and balsam is readily secured under this method of cutting if the following rules are observed:

1. In logging, the trees should be removed not singly but in small groups. The removal of such groups of trees will make small openings, or "holes," in the forest, which are more readily stocked than openings made by the removal of single trees. When single trees are cut, the openings are soon closed by the growth of side branches of the neighboring trees, and the young growth that appears is soon either shaded out or stunted. Openings, or "holes," in the forest formed by the removal of groups of trees a quarter of an acre or less in extent receive abundant seed from the surrounding trees, yet have enough light for a vigorous and normal development of the reproduction that springs up.

2. The same ground should not be logged too often; say, not oftener than every 10 or 20 years. Frequent logging over the same area prevents the firm establishment of young growth.

3. Keep out fires from the logged-over areas.

This system of gradual cutting, which may be called a selection system in groups, is decidedly the most practical, simplest, and safest so far as securing natural reproduction of spruce and balsam is concerned. Under it, spruce reproduction is favored at the expense of balsam, since the openings are small and the light conditions more favorable to spruce than to balsam. The greatest advantage of the system, however, is the protection which it affords against windfall—a very important consideration in all spruce cuttings.

The system differs from the method of logging practiced 25 to 30 years ago only in that the trees are cut in small groups instead of singly. Many of the old cuttings, when fires were kept out, have been cut over for the second and third time. Experience shows that no forest has ever been ruined by such a method of cutting. It is the recent logging, which amounts to practically clear cutting, especially when followed by fires, which has reduced large areas of timberland to a state where artificial planting or sowing is the only means of bringing them back to forest.

By clear cutting small groups, opportunity is afforded for utilizing all the merchantable timber, especially if the openings are made in the older and more mature stands. At the same time, forest conditions are preserved which are favorable for natural reproduction. The danger from windfall under this method is almost entirely avoided.

Cutting to a diameter limit.—Cutting in strips or selection cutting in groups requires a careful selection of the logging areas and expert technical knowledge. Wherever such knowledge can not be had, light cutting over the entire logging area may roughly answer the requirements of natural reproduction of both spruce and balsam fir. The higher the diameter limit for both species the more favorable will be the conditions for natural reproduction. The diameter limit should be raised in thin stands and lowered in dense ones, the main point being not to open the stand too heavily and destroy the conditions under which natural reproduction takes place. Although by cutting balsam fir to a lower diameter than spruce some advantage may be given spruce in reseeding the ground, yet under such a rough system it is difficult to control the conditions under which one or the other species can best come up; the preponderance of spruce or balsam fir in the future stand must therefore be left largely to chance.

ROTATION.

The difference in the rate of growth of balsam fir and spruce has a direct bearing upon the choice of rotation or proper time of cutting the two species. From the tables it is evident that balsam fir, if its growth is to be utilized to the fullest advantage, should not be cut before it reaches an age of about 100 or 125 years, or a diameter of 12 to 14 inches breast high. Cutting balsam fir below 6 or 7 inches means utilization of trees which are still making a fair growth. Spruce, on the other hand, should not be cut before it is 175 or 200 years old, since most of its growth is made at the age of from 100 to 200 years. The rotation for balsam fir, therefore, should be about 100 years, and for spruce at least 175 years. These rotations, of course, would be applicable only if balsam fir and spruce were grown separately. Since they usually grow together, the practical application of these different rotations would simply mean that in cutting over a virgin stand of spruce and balsam fir, the fir should be cut to a younger age, only the older spruce being removed.

SUMMARY.

1. Balsam fir forms, on an average, from 10 to 15 per cent of the entire red-spruce stand, or 5,355 million board feet.

2. Under present methods of cutting, balsam fir is increasing at the expense of red spruce in the second growth throughout the entire range of the two species.

3. Balsam-fir wood, while to some extent inferior to spruce for construction material, has a definite place in the pulp and lumber industries.

4. Balsam fir grows much faster throughout its entire life than spruce, but is shorter lived and reaches maturity long before the latter.

5. Balsam fir should be cut at an age of from 100 to 125 years, while spruce, as it grows at present in the wild wood, should be cut at an age of from 175 to 200 years.

6. The annual increment per acre of balsam fir throughout its range varies from one-sixth to one-third of a cord, or 1 cord in from three to six years.

7. The best silvicultural system of cutting is that of selection cutting in small groups. The natural reproduction of both spruce and balsam fir is assured under this system, with the possibility of increasing the proportion of spruce in the new stand.

BIBLIOGRAPHY.

ANDERSON, ALEXANDER P. Comparative Anatomy of the Normal Diseased Organs of Abies Balsamea, Affected with Aecidium Elatinum. (Botanical Gazette, Nove, pp. 1897, v. 24, pp. 309–344.)

Balsam Fir, il. (Hardwood Record, Apr. 25, 1908, v. 26, No. 1, pp. 16–17.)

" CLARK, J. F. On the Form of the Bole of the Balsam Fir. (Forestry Quarterly, Jan. 1903, v. 1, pp. 56–61.)

DORNER, HERMAN B. The Resin Ducts and Strengthening Cells of Abies and Picea. (Indiana Academy of Science, Proceedings, 1899, pp. 116–129.)

ENGELMANN, GEORGE. A Synopsis of the American Firs. (Transactions of the Academy of Science of St. Louis, 1878, v. 3, No. 4, pp. 593–602.)

HUNTINGTON, A. O. Balsam Fir. (New England Magazine, Oct. 1904, n. s. v. 31, p. 225.)

McADAM, T. The "Human Interest" in Firs. (Garden Magazine, Aug. 1909, v. 10, No. 1, pp. 12–14.)

MIYAKE, K. Contribution to the Fertilization and Embryogeny of Abies Balsamea. (Beihefte zum Botanische Centralblatt, 1903, v. 14, pp. 134–144.)

MOORE, B., and ROGERS, R. L. Notes on Balsam Fir. (Forestry Quarterly, March 1907, v. 5, pp. 41–50.)

ROTHROCK, J. T. Balsam Fir. (Forest Leaves, Feb. 1910, v. 12, No. 7, p. 105.)

VON SCHRENK, HERMANN. Glassy Fir. (Missouri Botanical Garden. 16th Annual Report, 1905, pp. 117–120.)

Lightning Source UK Ltd.
Milton Keynes UK
UKHW020624060119
334855UK00006B/385/P